ENGINEERING ECONOMICS FOR
PROFESSIONAL ENGINEERS' EXAMINATIONS

ENGINEERING ECONOMICS FOR PROFESSIONAL ENGINEERS' EXAMINATIONS

MAX KURTZ, P.E.

CONSULTING ENGINEER AND EDUCATOR; MEMBER,
NATIONAL SOCIETY OF PROFESSIONAL ENGINEERS;
AUTHOR, *Comprehensive Structural Design Guide,
Structural Engineering for Professional Engineers'
Examinations, Framing of Hip and Valley Rafters;*
EDITOR-IN-CHIEF, *Handbook of Engineering
Economics* (in preparation);
CONTRIBUTING AUTHOR, *Standard Handbook of
Engineering Calculations*

SECOND EDITION

McGRAW-HILL BOOK COMPANY

New York   St. Louis   San Francisco   Auckland   Düsseldorf
Johannesburg   Kuala Lumpur   London   Mexico   Montreal
New Delhi   Panama   Paris   São Paulo   Singapore
Sydney   Tokyo   Toronto

Library of Congress Cataloging in Publication Data

Kurtz, Max, date.
    Engineering economics for professional engineers'
examinations.

    Bibliography: p.
    Includes index.
    1. Engineering economy.   I. Title.
TA177.4.K87   1975   513'.93   75-2172
ISBN 0-07-035675-0

234567890   MUBP   78432109876

The editors for this book were Tyler G. Hicks and Margaret Lamb,
and the production supervisor was George E. Oechsner.
It was set in 8A by Bi-Comp, Incorporated.

Printed by The Murray Printing Company and bound by The Book Press.

TO RUTH

"For there shall be a sowing of peace and prosperity; the vine shall yield its fruit, and the ground shall give its increase, and the heavens shall give their dew. . . ."

ZECHARIAH 8:12

CONTENTS

PREFACE

Engineering transcends pure technology. The basic functions of engineering are to provide public services with maximum safety and at minimum cost, to enhance the pleasures of life, and to offer feasible solutions to the multitude of problems that continually arise in our complex industrial-social system. Thus, engineering design pertains to the real world rather than to some idealized world, and therefore, it must encompass the economic factors with which the real world is permeated. As a consequence, the study of economics is an integral part of the study of engineering, on a par with the study of beam behavior, heat transfer, and servomechanisms.

There are myriad ways in which economics influences engineering design, and only several of these can be cited at this point.

First, engineering design entails a choice among alternative schemes, for there are invariably multiple methods by which a required task can be performed. The engineer, therefore, must compare these alternative schemes with respect to both their technical efficiency and their cost. Generally, the sets of payments associated with the alternative schemes differ not only in amount but in timing as well. Therefore, in order to make a valid cost comparison of the alternative schemes, the engineer must have a clear grasp of the time value of money and be proficient in the mechanics of applying money-time relationships in his calculations.

Secondly, the magnitude of the project that the engineer designs is governed by the financial resources of the client. The engineer, therefore, must recognize the financial bounds within which he is constrained to operate. This matter is of particular importance in determining whether a project designed in anticipation of future growth should be constructed in one step or in multiple steps that correspond to the rate of growth.

Thirdly, a knowledge of engineering and finance is indispensable to the engineer who assumes the role of corporation executive. With the continuous expansion of our technology, industry is recruiting technically trained individuals into managerial positions to an ever-increasing extent. The highly technical nature of many of the products and services that industry offers makes it imperative that those who direct our commercial activity have an adequate understanding of the technology from which those products and services evolved.

In addition to pure economics, the engineer also requires a knowledge of probability theory. Engineering is oriented toward the future, for every engineering project must satisfy some social or industrial need for a specific period of time beyond the present. Consequently, engineering design is based on estimates, forecasts, and projections; it involves probability rather than certainty. Therefore, in the same way that engineering design must embody calculations of compound interest, it must also embody calculations of probability. Skill in applying the concepts, laws, and techniques of probability is of prime necessity to the design engineer.

This text has a threefold purpose. It is designed for use in a college course in engineering economics, for use by engineers preparing for the examinations for the professional engineers' license or a position in governmental service, and for use by practicing engineers and others concerned with engineering economics. With reference to the second item, this text covers the material that appears in the examination prepared by the National Council of Engineering Examiners and the examinations prepared by several individual states. To aid the candidate for licensure, the text includes a vast number of problems that have appeared in the professional engineers' examinations. They have been carefully selected as representative of P.E. examination problems, and they will afford to the candidate for licensure extensive practice in interpreting and solving examination problems.

Because many engineers encounter difficulty in grasping the terminology of finance, special emphasis has been placed on presenting all definitions in as simple and lucid a manner as possible. Moreover, since there are usually several alternative methods of solving a given problem, an attempt has been made in all cases to identify the method of solution that is simplest and most direct. The rule of continuity that is presented in the study of sinking funds and annuities illustrates this fact.

In Chap. 11, the subject of probability is developed in a very practical manner, with abstract theory held to a minimum. Considerable attention is devoted first to the subject of permutations and combinations since this constitutes the foundation of probabilistic analysis. Each problem in probability is viewed as a practical situation rather than as an exercise in applying abstract laws. Simple logic, rather than esoteric rules, is the keynote in this presentation. The probability of a specified event is determined by answering the following set of questions: How many outcomes are possible? Are these outcomes equally or unequally likely to occur? How many outcomes in the entire set will yield the specified event? These questions are answered by forming and counting permutations or combinations. The solution to Example 11-15 illustrates this simple analytical approach. Moreover, many problems in probability are solved by multiple methods, thus demonstrating the variety of ways in which each situation can be viewed and, it is hoped, making the subject more stimulating and meaningful to the reader.

Graphical representation is a potent tool in the study of any subject, thus this text makes prolific use of this expedient. Graphs are employed to represent such variable quantities as the principal of a loan, the principal in a sinking fund, the book value of a bond, etc. The advantages of graphical representation are numerous. A graph enables the viewer to perceive at a glance the data that are given and the information that is required; it depicts emphatically the manner in which a particular quantity varies in magnitude; it exhibits with remarkable clarity certain simple relationships that would otherwise remain hidden; finally, by clothing each problem in a tangible, geometrical form, a graph arouses the interest of the reader to a far greater extent than do numerical data alone.

As demonstrated in this text, it is possible in many instances to arrive at results in the subject of finance either by applying mathematics directly to the solution of a problem or by applying a rational method of pure reasoning, based on elementary economic considerations. This rational method is vested with a bare minimum of mathematics. Article 4-3, which presents an alternative derivation of the equation for the principal in a sinking fund, serves to illustrate the rational method of solution. This text applies the rational method more extensively than the mathematical method for the following reasons: the rational method is far simpler to apply, it leads us directly to conclusions that are not readily discernible through mathematical formulas alone, and it encourages us to think in terms of financial concepts and ideas.

Chapter 1 presents a review of the simple mathematical principles that underlie engineering economics. This material is also helpful as a review for the mathematics questions appearing in the P.E. examinations. While this text is concerned primarily with pecuniary considerations, we must not allow concentration on one subject to distort our perspective. In every situation there are inherent considerations that, although not susceptible to precise measurement in monetary units, nevertheless require careful evaluation in weighing the feasibility of a proposed course of action. For example, in selecting the site of a new factory, the matter of availability of skilled labor may outweigh the matter of cost. Similarly, the adoption of a business policy inimical to the public interest will eventually redound to the detriment of the enterprise. The influence of nonpecuniary factors must not be overlooked or minimized.

The second edition of this book reflects the strides that have been made in engineering economics, and it meets the requirements of current P.E. examinations. The major revisions are these:

1. The following material has been added: Art. 8-5 on linear programming for two variables, Chap. 10 on the elements of statistics, and Chap. 11 on probability.

2. The following material has been deleted: index numbers, professional ethics, and commercial and legal phases of the construction industry. Those who wish to read this material should refer to the first edition.

3. Much of the remaining material has been expanded.

In teaching P.E. review courses, the author has found that engineers are intrigued by the subject of engineering economics. Many engineers are leaders in their communities, and they eagerly seek a broad understanding of the manner in which economics and finance shape our society.

Max Kurtz

RECOMMENDATIONS TO THE READER

To gain a true understanding of engineering economics, we must become thoroughly familiar with its specialized vocabulary. Each term in this subject has a very precise technical meaning, one that may differ widely from its meaning in ordinary usage. Therefore, it is recommended that in studying this text you place strong emphasis on learning each definition until it becomes deeply embedded in your knowledge.

The reader who is preparing for a specific examination will find that solving a vast number of specimen problems is a very effective form of preparation. Considerable benefit will accrue if you solve all problems presented at the end of each chapter. With respect to problems in finance, it is always possible to test the accuracy of a numerical solution by evaluating all sums of money at some reference date other than the one used in achieving the solution. For example, if a problem pertaining to an annuity has been solved by selecting the origin date of the annuity as the reference date, the answer can be tested by transferring the reference date to the terminal date of the annuity, or to any other convenient date. After solving a problem to which the text does not supply the answer, verify your solution by applying this simple technique.

This text uses graphs extensively to assist the reader in visualizing each type of problem. Wherever feasible, construct a line diagram to represent the data in a given problem before you undertake the solution. This need not be a formal drawing; a freehand sketch will do. In fact, there is an advantage in exaggerating the curvature of each line in order to bring the inherent relationships into sharp focus. In an examination problem that requires a written solution, a simple drawing of this type constitutes prima-facie evidence that the examinee clearly understands the problem, and it consumes only seconds of his time.

In formulating the relationships and equations of finance, this text develops a rational method, as distinguished from the purely mathematical method. This matter is discussed in Art. 7-13, and the rational method is applied throughout the text. Seek additional applications of the rational method, either as an alternative means of formulating relationships that are presented in this text or as a means of discovering new relationships. This practice will enliven your study of engineering economics and enable you to develop a much deeper understanding of the subject.

This text exhibits various analogies that exist between situations in engineering economics and situations in engineering and science. For example, Art. 9-4 demonstrates that the continuous compounding of interest is analogous to many natural phenomena, and Art. 11-10 points out that the probability-density function is analogous to entropy. As you read this text, be alert to find other analogies. The discovery of analogies is exciting, and a recognition of analogies results in economy of effort. The theorems and equations of one situation can be transferred bodily to an analogous situation merely by replacing one set of terms and symbols with another. Duplication is eliminated entirely.

It is hoped that you will find the study of this book a stimulating and rewarding experience.

ENGINEERING ECONOMICS FOR
PROFESSIONAL ENGINEERS' EXAMINATIONS

CHAPTER 1

MATHEMATICAL REVIEW

1-1. Laws of Exponents. The expression a^n, read "a exponent n" or "a raised to the nth power," denotes the product obtained by multiplying the quantity a by itself n times. Thus, $8^3 = 8 \cdot 8 \cdot 8 = 512$. The basic laws of exponents are listed below:

1.
$$a^m a^n = a^{m+n} \tag{1-1}$$

This arises from the fact that multiplication is commutative; i.e., the order of multiplication is immaterial. For example

$$2^3 \cdot 2^4 = (2 \cdot 2 \cdot 2)(2 \cdot 2 \cdot 2 \cdot 2) = 2^7 = 128$$

2.
$$\frac{a^m}{a^n} = a^{m-n} \tag{1-2}$$

For example

$$\frac{2^5}{2^3} = \frac{2 \cdot 2 \cdot 2 \cdot 2 \cdot 2}{2 \cdot 2 \cdot 2} = 2^2 = 4$$

3.
$$a^{-m} = \frac{1}{a^m} \tag{1-3}$$

Proof. If k is any quantity,

$$\frac{a^k}{a^{k+m}} = \frac{a^k}{a^k a^m} = \frac{1}{a^m}$$

However, applying the second law, we obtain

$$\frac{a^k}{a^{k+m}} = a^{k-(k+m)} = a^{-m}$$

$$\therefore a^{-m} = \frac{1}{a^m}$$

4.
$$(a^m)^n = a^{mn} \tag{1-4}$$

1

Proof

$$(a^m)^n = a^m \cdot a^m \cdot a^m \ \cdots \ n \text{ times}$$
$$= a^{m+m+m+} \cdots = a^{mn}$$

5. $$(ab)^n = a^n b^n \tag{1-5}$$

Proof

$$(ab)^n = (ab)(ab)(ab) \ \cdots \ n \text{ times}$$
$$= (a \cdot a \cdot a \ \cdots \ n \text{ times})(b \cdot b \cdot b \ \cdots \ n \text{ times})$$
$$= a^n b^n$$

1-2. Logarithms. If $a^y = n$, then y is said to be the logarithm of n to the base a. In other words, the logarithm of a given number to a given base represents the power to which the base number must be raised to obtain the given number. A logarithm is therefore an exponent. Thus, since $7^3 = 343$, we can say that the logarithm of 343 to the base 7 is 3. The following are the basic rules of logarithms:

1. $$\log mn = \log m + \log n \tag{1-6}$$

Proof. Let $u = \log m$ to base a, and $v = \log n$ to base a.

Then $$a^u = m \quad \text{and} \quad a^v = n$$
$$\therefore mn = a^u a^v = a^{u+v}$$
$$\therefore \log mn = u + v = \log m + \log n$$

2. $$\log \frac{m}{n} = \log m - \log n \tag{1-7}$$

Proof

$$\frac{m}{n} = \frac{a^u}{a^v} = a^{u-v}$$

$$\therefore \log \frac{m}{n} = u - v = \log m - \log n$$

3. $$\log m^n = n \log m \tag{1-8}$$

Proof

$$m^n = m \cdot m \cdot m \ \cdots \ n \text{ times}$$
$$\therefore \log m^n = \log m + \log m + \ \cdots \ n \text{ times}$$
$$= n \log m$$

Although this proof is restricted to positive integral values of n, the law applies for all values of n.

In our logarithmic calculations, we will use the so-called "common" system of logarithms, having 10 as a base. Hence, the log of 10 is 1, the log of 100 is 2, etc. From tables, we find that

$$10^{0.3096} = 2.04$$
$$\therefore \log 2.04 = 0.3096$$

If we multiply 2.04 successively by 10 and apply the first logarithmic rule, we obtain the following results:

$$\log 2.04 = 0.3096$$
$$\log 20.4 = \log (2.04 \times 10) = \log 2.04 + \log 10 = 1.3096$$
$$\log 204.0 = \log (2.04 \times 100) = \log 2.04 + \log 100 = 2.3096$$

The integral portion of a logarithm is referred to as the "characteristic," while the fractional part is called the "mantissa." Thus, the logarithm of 204.0 has a characteristic of 2 and a mantissa of 0.3096. From the preceding results, the following rules can be formulated:

1. For a given sequence of integers, the mantissa of the logarithm is independent of the position of the decimal point.

2. To determine the characteristic of the logarithm of a number greater than unity, subtract 1 from the number of digits to the left of the decimal point.

Logarithmic tables list only the mantissa of a given number; the characteristic is then obtained by applying the rule stated above. For example, assume we are to find the logarithm of 37.2. From Table A-1 of the Appendix, the mantissa is 0.5705; since there are 2 digits to the left of the decimal point, the characteristic is 1. Hence, log 37.2 = 1.5705. Conversely, if we are to find the number corresponding to a given logarithm, we obtain from the table the sequence of integers having the given mantissa and then locate the decimal point according to the given characteristic.

Although a number less than unity has a negative logarithm, it is convenient to express its logarithm as the algebraic sum of a positive mantissa and a negative characteristic. In this manner, a given sequence of integers retains the same mantissa regardless of the location of the decimal point. Moreover, the negative characteristic is expressed by subtracting 10 from a lesser number. To illustrate the use of this convention, we shall divide 2.04 successively by 10 and apply the second logarithmic rule.

$$\log 2.04 = 0.3096$$
$$\log 0.204 = \log \frac{2.04}{10} = \log 2.04 - \log 10$$
$$= 0.3096 - 1 = 9.3096 - 10$$
$$\log 0.0204 = \log \frac{2.04}{100} = \log 2.04 - \log 100$$
$$= 0.3096 - 2 = 8.3096 - 10$$
$$\log 0.00204 = \log \frac{2.04}{1000} = \log 2.04 - \log 1000$$
$$= 0.3096 - 3 = 7.3096 - 10$$

From the foregoing results, we can deduce the following rule: To determine the characteristic of the logarithm of a number less than unity, subtract from 9 the number of ciphers between the decimal point and the first significant figure. From the number thus obtained subtract 10.

The following numerical examples will serve to illustrate the method of applying logarithms to arithmetical calculations:

1. Find 529×62.4.

Solution

$$\log (529 \times 62.4) = \log 529 + \log 62.4$$
$$= 2.7235 + 1.7952 = 4.5187$$
$$\therefore 529 \times 62.4 = 33,010 \qquad \text{to four significant figures}$$

2. Find $(11.3)^{2.5}$.

Solution

$$\log (11.3)^{2.5} = 2.5 \times \log 11.3$$
$$= 2.5 \times 1.0531 = 2.6328$$
$$\therefore (11.3)^{2.5} = 429.3$$

1-3. Arithmetical Series. An "arithmetical series" is a sequence of terms in which each term differs from the preceding one by a fixed amount called the common difference. Thus, 2, 8, 14, 20, 26, . . . is an arithmetical series having the common difference 6.

1-4. Geometrical Series. A "geometrical series" is a sequence of terms in which the ratio of each term to the preceding one is a constant. Thus, 2, 6, 18, 54, 162, . . . is a geometrical series having the ratio 3. We shall now derive an equation for the sum of the first n terms of a geometrical series. In general, let

$$S_n = a + ar + ar^2 + \cdots + ar^{n-2} + ar^{n-1}$$

Multiplying each member of this equation by r, we obtain

$$S_n r = ar + ar^2 + ar^3 + \cdots + ar^{n-1} + ar^n$$

Subtracting the first equation from the second,

$$S_n(r - 1) = ar^n - a = a(r^n - 1)$$

$$\therefore S_n = a\frac{r^n - 1}{r - 1} \qquad (1\text{-}9)$$

1-5. Binomial Theorem. The expression $n!$, read "factorial n," is used to denote the product of the first n positive integers. Thus,

$$5! = 1 \cdot 2 \cdot 3 \cdot 4 \cdot 5 = 120$$

If the binomial $(a + b)$ is raised to the nth power, we obtain through successive multiplication

$$(a + b)^n = a^n + \frac{n}{1!} a^{n-1}b + \frac{n(n - 1)}{2!} a^{n-2}b^2$$

$$+ \frac{n(n - 1)(n - 2)}{3!} a^{n-3}b^3 + \cdots \qquad (1\text{-}10)$$

1-6. The Derivative of a Function. If y is a function of x, then the ratio of the rate of change of y to the rate of change of x is termed the "derivative" of y with respect to x, and is written dy/dx.

In texts on calculus, the following equations for derivatives are derived:

Equation	*Derivative*	
$y = ax$	$\dfrac{dy}{dx} = a$	$(1\text{-}11)$
$y = x^n$	$\dfrac{dy}{dx} = nx^{n-1}$	$(1\text{-}12)$
$y = uv$	$\dfrac{dy}{dx} = u\dfrac{dv}{dx} + v\dfrac{du}{dx}$	$(1\text{-}13)$
(where u and v are functions of x)		
$y = \dfrac{u}{v}$	$\dfrac{dy}{dx} = \dfrac{v\dfrac{du}{dx} - u\dfrac{dv}{dx}}{v^2}$	$(1\text{-}14)$

Figure 1-1 is a graph of the equation by which y is related to x. Geometrically, the ratio of the rate of change of y to the rate of change of x is

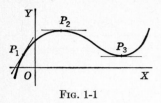

FIG. 1-1

equal to the slope of the curve at the particular point in question, and the slope of the curve is measured by the slope of the straight line that is tangent to the curve at that point. Hence, the derivative of y with respect to x equals the slope of the tangent line. For example, if the tangent at point P_1 has a slope of 2, the derivative at this point equals 2; y is increasing twice as rapidly as x.

At point P_2, y attains its maximum value for that region, and at P_3 it possesses its minimum value for this other region. At both these points, the tangent line has zero slope. Therefore, if we are required to determine the points at which a given function has its maximum or minimum value, it is simply necessary to equate its derivative to zero. In problems of this nature that we will encounter in this text, it will be evident from the conditions of the problem whether the point in question is one of maximum or minimum value.

1-7. Properties of a Straight Line. Consider that two variables, x and y, are related to one another by the equation

$$ax + by = c_1$$

where a, b, and c_1 are positive constants. The graph of this equation is the straight line L_1 in Fig. 1-2.

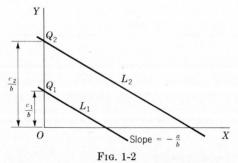

FIG. 1-2

The slope of L_1 is found in this manner: Solving for y,

$$y = \frac{c_1}{b} - \frac{a}{b}x$$

Then
$$\text{slope of } L_1 = \frac{dy}{dx} = -\frac{a}{b} \qquad (1\text{-}15)$$

The slope of the line is therefore independent of c_1.

The ordinate (y value) of the point Q_1 at which this line intersects the YY axis is called the "y intercept" of the line. Setting $x = 0$, we obtain

$$y \text{ intercept of } L_1 = \frac{c_1}{b} \qquad (1\text{-}16)$$

Since b and c_1 have the same algebraic sign, the y intercept is positive, and L_1 intersects the YY axis above the origin.

Now consider that c_1 is increased to a higher value c_2, giving

$$ax + by = c_2$$

The graph of the new equation is the straight line L_2 in Fig. 1-2. In accordance with Eq. (1-15), L_2 is parallel to L_1. In accordance with Eq. (1-16), L_2 lies at a greater distance from the origin than does L_1.

In summary, let us say that x and y are related to one another by the equation

$$ax + by = c$$

where a and b are positive constants and c is a variable that is restricted to positive values. If we start with one value of c and then assign successively higher values to c, the straight line L that represents the given equation is displaced so that:
1. It remains parallel to its original position.
2. Its distance from the origin increases.

PROBLEMS

1-1. What is the common logarithm of $1.67^{2.3} \times 0.401^{0.6}$?   *Ans.* 0.2741

1-2. The following are the first four terms of a series formed according to a definite rule: 2, -6, 18, -54. Find the sum of the first nine terms of this series.
Ans. 9,842

1-3. What is the eighth term in the expansion of $(a + b)^{12}$?   *Ans.* $792a^5b^7$

1-4. The volume of a sphere is increasing at the rate of 6 cu in. per sec. Find the rate at which the radius is increasing when the volume is 21 cu in.   *Ans.* 0.163 ips

1-5. A farmer wishes to fence a rectangular field on three sides. He has 160 rods of wire fencing. What should be the dimensions of the field to yield the maximum area?   *Ans.* 40 by 80 rods

CHAPTER 2

THE MECHANICS OF INVESTMENT

2-1. The Interest-earning Property of Money. If an individual owns some tangible asset which he does not require for his own use, he can exploit this asset as a source of income by renting it to others. The asset can be an apartment house, a parcel of land, or a warehouse; the sole requirement is that a demand for the asset shall exist. The amount of rent which the owner can obtain will depend upon such factors as the demand for this particular asset, the cost of maintaining the asset in operating condition, and the rapidity with which the asset depreciates in value. Thus, we see that a physical asset, in addition to its actual present value, possesses a latent future value arising from its availability for rental.

Similarly, if an individual possesses a sum of money in excess of his immediate requirements, he can utilize this excess money as a source of income by lending it to others. In this case also, the sole requirement is that a demand for borrowed money shall actually exist. The sum of money loaned is designated the "principal" of the loan, and the charge which the borrower pays to the lender for the use of his money is termed the "interest." The ratio of the interest payment to the principal for a given unit of time is termed the "interest rate" and is usually expressed as a percentage of the principal. Thus, if the sum of $1,000 is borrowed and the debtor is obligated to pay the creditor $60 for each year the loan is in existence, then

$$\text{Interest rate} = \frac{\text{interest paid per annum}}{\text{principal}}$$

$$= \frac{\$60}{\$1,000}$$

or $\qquad$ Interest rate $= 6$ per cent per annum

8

We have defined interest as being essentially the compensation offered for the use of borrowed money. When used in this restricted sense, the interest attaching to a loan is termed "pure interest." The pure-interest rate which a given sum of money can command will depend upon such factors as the demand that exists for borrowed money, the supply of money available for lending, and the magnitude of this particular sum. (In general, a large sum of money can be employed more profitably for lending purposes than a comparatively small sum.)

There are, however, certain additional factors that determine the charge to be paid for borrowed money, such as the administrative and legal expenses incurred by the creditor in granting a loan or the degree of risk assumed by the creditor in entrusting his money to the debtor. Since these factors are always present, pure interest alone is a somewhat abstract concept. No attempt is generally made to distinguish these component elements in determination of interest from one another, and the term "interest" is employed to denote their combined result.

Thus, in the same way that a physical asset possesses latent future value by virtue of its rental possibilities, so too a given sum of money is vested with latent future value through its availability for lending.

As developed above, the interest-earning capacity of money applies only to surplus money not required by its owner for his immediate use. Viewed from the reciprocal relationship of the debtor, a sum of money which he is compelled to borrow carries with it the obligation to pay interest. In our study of finance, however, it is necessary to recognize the interest-earning (or interest-paying) property of money as a characteristic applying to every sum of money, regardless of the manner of its disposal. It is irrelevant to us whether this sum constitutes the principal of a loan, is expended to make a necessary purchase, or is simply hoarded. Since the possibility of earning interest exists, that possibility requires consideration in the making of any financial decision.

In order to illustrate this point, let us assume that a business firm is considering the feasibility of purchasing the building in which it operates and which it is currently renting. Assume first that the firm must borrow the funds necessary to purchase the building. It must pay interest until the debt is discharged, and this interest is obviously a direct cost of owning the building. Now assume that the firm can purchase the building with its own funds. By devoting its capital to this purchase, the firm will divert capital from other channels in which it is capable of earning interest.

It thus follows that purchase of the building entails forfeiture of potential interest income, a factor that requires recognition in arriving at a decision. This can be taken into account by including the lost income in the cost of owning the building.

In the first instance, where money was borrowed, interest existed as a direct expense; in the second instance, where no borrowing occurred, interest entered into the calculations in the form of forfeited income. Although the specific amounts of interest involved in the two cases would probably not be identical, the end results are essentially the same.

2-2. The Need of Borrowed Money. Since it is the need of borrowed money that endows money with its interest-earning capacity, we shall now investigate the basic factors that produce and sustain that need. In our complex industrial system, there are myriad forces in operation that impel the borrowing of money. The following are of paramount importance:

Capital Expansion. All commodities that are produced by industry can be classified into two broad categories, namely, consumer goods and capital goods. "Consumer goods" are those which are used directly by the general public; all other commodities are referred to as "capital goods." A capital commodity can be employed either to produce consumer goods or to produce other capital goods. Thus, an automobile used for pleasure driving is consumer goods, as is also the gasoline it consumes. On the other hand, a machine that manufactures the component parts of the automobile is capital goods, as is also the factory in which the machine is located. The characteristic that renders a commodity either consumer or capital goods is not necessarily an innate property of the commodity itself; the criterion is simply the mode of its consumption. For example, fuel oil is consumer goods when used to heat a residential building, but it is capital goods when used in a factory.

Now, under a highly dynamic industrial system such as ours, there occurs a continuous expansion of the volume of existing capital goods. The primary factors stimulating this expansion are the following:

TECHNOLOGICAL IMPROVEMENTS. When a product is invented that is of great benefit to the public, new capital goods must be created before that product can be placed on the market for distribution to the public. There must come into existence new machinery to manufacture the product, new factories in which to perform the manufacture, new warehouses in which to store the product, etc.

AN INCREASE OF POPULATION. If we assume that the volume of capital goods required to maintain a particular standard of living is approximately proportional to a nation's population, it follows as a corollary that an increase of population must be accompanied by a corresponding increase of capital commodities.

An expansion of the volume of capital goods can only be accomplished through the expenditure of vast sums of capital. Moreover, a considerable period of time will elapse before the financial returns obtained from the new commodities will be sufficient to return to the owners the capital originally expended. To illustrate this, let us consider the case of a business firm constructing a factory in order to manufacture a newly invented product. The factory must be paid for as it is constructed, but the income derived from its use, in the form of profits resulting from the sale of the product, will be obtained gradually. Should the firm lack the necessary resources, it is compelled to borrow money to finance the undertaking. Thus, the need for capital expansion contributes very substantially to the demand for borrowed money.

Evidently, the financial returns obtained through capital expansion must be sufficient to retrieve the invested capital within a reasonable period of time, to permit the payment of interest for the money borrowed, and to provide an income for the business enterprisers.

Expansion of Public Services. With the continuously altering pattern of our social and industrial organization, it becomes necessary for all governmental bodies to expand constantly the services and facilities they provide for the public. The development of automotive transportation has required the construction of new highways; the increase of population has required the construction of new schools and public hospitals; the growth of the suburbs has required the construction of new waterworks, the paving of new streets, etc. Here again, if the governmental body involved has insufficient resources to finance a particular project, it must resort to borrowing.

The process of borrowing on the part of a governmental division is generally implemented by the issuance of bonds, which will be studied in detail at a later point. The revenue necessary to redeem the bonds at their maturity can be accumulated either through general or special taxation, or through direct charges on those using a particular facility. Illustrations of the latter form of obtaining revenue include tolls on state highways, charges for municipally owned waterworks, etc.

Working Capital for Business. Business firms are often obliged to borrow funds in order to augment their inventories, expand their facilities, or tide themselves over a slack period.

Expansion of Consumer Commodities. In recent times, consumer credit has come to play an increasingly prominent role in the economy of this nation. The development of many products that contribute toward the enjoyment of life, but whose initial cost is beyond the financial means of many families, has stimulated the widespread use of installment buying, and the steadily increasing degree of home ownership has increased the volume of mortgage financing.

In explaining the nature and function of interest, classical economists have emphasized almost exclusively the role played by capital expansion. Interest is regarded as an inducement offered to the public to abstain from consuming its entire income in order that a portion of this income shall be available for the creation of new capital commodities. The possibility of paying interest is then ascribed to the fact that the wealth obtained through capital expansion is of such magnitude as to permit a modest reward to the creditors who made this expansion possible.

This limited interpretation of the economic significance of interest may have sufficed at a time when our industrial and agricultural economies were comparatively simple and when governmental operations were far less extensive than at the present time. A proper understanding of the role of interest in our present civilization, however, requires the inclusion of all the factors enumerated above.

2-3. The Media of Investment. The process of lending money for the purpose of earning interest is referred to as "investment." When used in this sense, the term implies a complete relinquishment of the use of the principal on the part of the lender. Stuart Chase has defined investment as "spending by proxy." If an entrepreneur employs his capital toward an expansion of his own business, such use is not deemed an investment, since the act of lending money is absent from this situation. However, where an individual purchases corporate stock for the primary or exclusive purpose of securing dividends, an investment is considered to have been made, since the individual is in effect granting the corporation the use of his capital. The fact that the stockholder is theoretically a part owner of the corporation is disregarded if his share of the total ownership is minute, or if he has no intention of assuming an active part in the management of the corporation.

The limitation of this definition is that it fails to take cognizance of many borderline cases, in which an individual assigns capital to a business enterprise and thereby obtains a substantial portion of the ownership of the firm.   Moreover, in our study of finance, we are concerned solely with the fact that money possesses a growth potential, irrespective of the precise manner in which that potential growth is converted to reality. Therefore, we shall apply the term "investment" in the broadest possible sense, that is, to denote any productive employment of capital.

Our next problem is to consider the various channels through which capital flows from the investor to the point at which it is expended.   We shall consider in turn the most important media that exist for the investment of capital, which are the following:

1. The purchase of corporate stock
2. The granting of direct loans
3. The purchase of a share in a savings and loan association
4. Various indirect forms of investment

2-4. The Corporate Structure.   A "corporation" is an aggregation of individuals formed for the purpose of conducting a business and recognized by law as a fictitious person.   Ownership of a corporation is represented by capital stock, which is divided into units called "shares."   The profits earned by a corporation are periodically distributed to the stockholders in the form of "dividends."   Two important advantages are inherent in the corporate form of business organization: first, the stockholders are not personally liable for the debts of the firm; second, since a share of stock is negotiable, the continuity of the enterprise is not affected by the death of a stockholder.

A corporation is chartered by the state in which it conducts its business, or in which it maintains its central office.   The "charter," or "certificate of incorporation," states among other things the name of the corporation, the maximum number of shares which it is authorized to issue, and the initial amount of capital.   The business is conducted by a board of directors elected by the stockholders.

Although most corporations issue only one type of stock, there are some whose stock is divided into two categories, "common" and "preferred." While details vary, preferred stock generally has a prior claim to receive a fixed amount of dividends, with the residue of the dividends being available for the common stockholders.   In many instances, the preferred stock is vested with a cumulative right to dividends; that is, if the dividends for a particular period are not paid at the termination of that period, the claim of the preferred stockholders continues in force.   When dividends are declared at a future date, these accrued dividends must first be paid to the preferred stockholders before any may be paid to the common stockholders.   Where preferred stock is noncumulative, this accrual feature is not present; dividends that are passed at the end of a particular period are permanently lost.

The stockholders, as owners of the corporation, possess the right to vote at stockholders' meetings, to receive dividends when declared, to inspect the corporate books and records, and, in the event of dissolution, to share in the remaining assets.   With regard to the last-mentioned right, the preferred stock often bears a claim prior to that of the common stock. Stockholders have the right to arrest any *ultra vires* act on the part of the corporation (i.e., one not specifically authorized by the corporate charter), but such an act is difficult to define in view of the broad nature of most charters.

Stockholders of a corporation also possess a right to maintain unimpaired their proportionate share of the ownership of the concern.   In deference to this right, a corporation that is increasing its volume of outstanding stock will offer the present stockholders the privilege of subscribing to the new issue of stock at a price substantially below the prevailing market price of the old stock and in a proportion based upon the number of shares already owned by each stockholder.   For example, assume that there are 1,000 shares of stock outstanding and that the corporation is issuing 200 new shares.   Each share of stock possesses a "right" to subscribe to the new shares; it will thus require 5 rights to purchase 1 new share.   A stockholder owning 40 shares acquires 40 rights, enabling him to subscribe to 8 new shares.

The corporation issuing new stock will announce in advance that the rights will be granted to all stockholders of a specified record date and will stipulate the date at which this option expires.   Stock rights are evidenced by certificates called "warrants."   These warrants are negotiable instruments, and a stockholder not wishing to exercise his option can sell his rights.   The monetary value of a stock right resides in the fact that it permits the purchase of new stock at a price appreciably below the prevailing market price of the original stock.

The sale of new stock at a lower price will tend to depress the value of the old stock. To understand the underlying cause, let us assume that a corporation has 1,000 shares of stock outstanding, whose market value is $100 per share, and that it offers to its stockholders 200 new shares at a subscription price of $88 each. Under the original circumstances, the total theoretical value of the 1,000 outstanding shares of stock is $100,000. After the sale of 200 new shares at $88 each has been consummated, the corporation has increased its capital by $17,600. Hence, there are now 1,200 shares outstanding, whose total theoretical value is $117,600, or $98 per share. If we assume that all other factors remain constant, the latter sum will represent the new market value of a share of stock subsequent to the record date, when the stock is divested of its stock rights. The theoretical value of the stock right is therefore $2.

Thus, during the interim between the record date and the expiration date of the stock right, a new investor can acquire one new share of stock by purchasing 5 rights at $2 each and applying these rights to the purchase of a new share for $88, a total expenditure of $98. An investor who sells his stock right for $2 is receiving compensation for the decline in the value of his stock. In general, let

MP = market price of a share of stock immediately prior to the record date

SP = subscription price of a new share

VR = theoretical value of one stock right

n = number of rights required to purchase one new share

Then

$$VR = MP - \frac{n(MP) + SP}{n + 1}$$

or

$$VR = \frac{MP - SP}{n + 1} \tag{2-1}$$

2-5. Promissory Notes. The next medium of investment to be considered is the granting of direct loans by one party to another. A loan whose duration is 5 years or less is generally referred to as a "short-term loan," and one of greater duration, as a "long-term loan." Short-term loans are usually evidenced by promissory notes.

A "promissory note" is an unconditional written promise made by one party, termed the "maker," to another, to pay a stipulated sum of money either on demand or at a definite future date. The sum of money stated in the note is called its "face value," and if the note is payable at a definite date, this date is termed its "due date" or "date of maturity." If no mention of interest appears, the maturity value of the note is its face value; where the payment of interest is specified, the maturity value is the face value of the note plus the accrued interest.

For example, assume that a promissory note dated Jan. 10, 1975, reads as follows: "Three months after date I promise to pay to the order of Henry Johnson the sum of $1,000, with interest at 6 per cent per annum, at the First National Bank, Canton, Ohio."

The face value of the note is $1,000; its due date, or maturity date, is Apr. 10, 1975; and its maturity value is calculated in the following manner:

$$\text{Interest} = \$1,000 \times 0.06 \times \tfrac{1}{4} = \$15$$
$$\therefore \text{Maturity value} = \$1,015$$

2-6. Mortgage Loans. In the case of many types of loans, regardless of the duration, the creditor will require some form of security for the repayment of the loan, such as property owned by the debtor. The legal instrument under which the property is pledged is termed a "mortgage," or "trust deed." In general, title and possession of the property remain with the mortgagor, or borrower. However, since the mortgagee, or lender, has a vested interest in maintaining the value of this property unimpaired, at least to the extent of the unpaid balance of the loan, the terms of the mortgage generally require that the mortgagor keep the property adequately insured, make necessary repairs, and pay all taxes as they become due.

Should it occur that there is more than one lien applying to a particular piece of property, then the mortgages are referred to as the first mortgage, second mortgage, etc., corresponding to the priority of the claim. In the event of default in repaying the loan, or in the event of a violation of the terms of the mortgage on the part of the owner, the mortgagee can request a court of law to undertake the sale of the property in order to recover the balance of his loan.

Property is often purchased with borrowed funds, and the property thus acquired constitutes the security behind the loan. This is frequently true, for example, in the purchase of homes. Mortgage loans of this type generally possess an amortization feature. Although the term "amortization" means simply the liquidation (literally, "killing") of a debt, it is generally used to denote the process of gradually extinguishing a debt through a series of equal periodic payments extending over a stipulated period of time. Each payment can be regarded as consisting of a payment of the interest accrued for that particular period on the unpaid balance of the loan and a partial payment of principal. With each successive payment, the interest charge diminishes, thereby accelerating the reduction of the debt.

2-7. Bonds. Long-term obligations are usually evidenced by "bonds," which are promissory notes issued under seal by both corporations and governmental bodies. They are negotiable instruments. The legal document stipulating the terms of the bonds is known as the "bond indenture." The sum of money specified in the bond is termed the face, or "par," value. Most bonds provide for the payment of interest at a stipulated rate at stated intervals, with the principal to be paid to the bondholders when the bonds are redeemed at their maturity. For example, if a corporation issues a $1,000 bond maturing Dec. 31, 1990, with interest at 5 per cent payable semiannually, the bondholder will receive an interest payment of $25 at the expiration of each semiannual period and will receive the par value of $1,000 when the bond matures at the end of 1990. The capital obtained by a corporation through the issuance of bonds is materially reduced by various legal and administrative expenses incidental to the printing and sale of the bonds, the periodic payment of interest, etc.

Bonds which have no security behind them other than the assets of the corporation issuing them are known as "debenture bonds," while bonds which are secured by a lien on some specific property owned by the corporation are known as "mortgage bonds." It does not necessarily follow that bonds lacking specific security carry greater risk than those having such security; the debenture bonds of a strong corporation may constitute a safer investment than the mortgage bonds of a weak corporation.

As an assurance to the bondholders that the issuing corporation will accumulate the funds necessary to retire the bonds at their maturity, the bond indenture will often require that the firm reserve a fixed amount of its annual earnings for bond redemption, the sums thus held in reserve to be deposited in an interest-earning fund known as a "sinking fund." In some cases, this fund is to be established immediately upon issue of the bonds, while in other cases its creation is deferred until a specified future date. The principal to be accumulated in the sinking fund at the maturity date of the bonds, which will represent the total of the periodic deposits and the accrued interest, must equal either the full maturity value of the bonds to be redeemed or some stipulated fractional part thereof. Failure on the part of the corporation to make the required deposits in the fund constitutes default of the bond agreement. Bonds whose terms oblige the corporation to establish such a sinking fund are termed "sinking-fund bonds."

"Serial bonds" are bonds which are issued simultaneously but which mature at different dates. For example, a corporation may issue 1,000 bonds at the present date, 330 of which mature in 1995, 330 in 1996, and the remainder in 1997. Serial bonds are issued in cases where no sinking fund is established for the retirement of the bonds. By allowing them to mature in series, the corporation will ease somewhat the burden of accumulating the capital required for redeeming the bonds.

"Perpetual bonds" are those which have no definite maturity date; they therefore represent permanent obligations, and the periodic payment of interest continues indefinitely. Such bonds have been issued by the British government. In addition, there are certain bonds outstanding at present whose maturity dates lie so distant in the future as to warrant classifying them as perpetual bonds for all practical purposes.

A purchaser of corporate bonds is simply a creditor of the corporation; unlike a stockholder, he holds no equity in the concern and possesses no voting rights. However, some corporations issue bonds which empower the bondholder, at his option, to convert his bonds to corporation stock during a specified period of time. Such bonds are designated "convertible bonds." The bond indenture stipulates the conversion ratio pertaining to these bonds, i.e., the number of shares of stock that can be obtained in exchange for one bond.

An important characteristic of bonds issued by municipalities is that the interest income obtained by the bondholder is exempt from the Federal income tax. For this reason, municipal bonds represent an attractive form of investment for individuals in high income tax brackets but are unsuitable for other investors. For example, assume that an individual is subject to a Federal tax amounting to 50 per cent of his income. Evidently, the receipt of $20 interest on a tax-exempt municipal bond is equivalent to the receipt of $40 interest on a nonexempt bond, as far as the income remaining after the payment of Federal income tax is concerned. (The bond interest may, however, be subject to a state income tax and, in the event of the death of the bondholder, to a Federal estate tax.) It is this tax-exemption feature that explains why municipal bonds generally yield low interest rates, and their purchase is confined primarily to wealthy investors.

When the selling price of a bond coincides with its face value, it is said to be sold "at par"; when the selling price exceeds the face value, the bond is sold "at a premium"; finally, when the selling price is below the face value, it is sold "at a discount." The differential between selling price and face value is termed the "premium" or "discount," whichever applies. If a bond is purchased at a premium or discount, the true interest rate earned by the investor differs from the interest rate quoted in the bond. To illustrate, assume that an investor purchases for $900 a $1,000 bond paying interest at 5 per cent annually. Not only is it true that the $50 annual dividend represents more than 5 per cent of his initial investment, but at maturity the investor will receive $100 in excess of his investment. For every purchase price, there is a corresponding true interest rate, and vice versa. Hence, if a prospective investor wishes to obtain an investment rate different from the nominal interest rate of the bond, he can make the necessary adjustment by offering to purchase the bond at the price corresponding to the intended rate.

There are many factors that determine the selling price of a bond (or the investment rate the bond yields, since one is a function of the other). Several of these are the following:

1. The degree of risk inherent in the purchase of the bond, as evidenced by the credit standing of the issuer. In general, the greater the risk, the greater will be the investment rate which the purchasers will seek.

2. The conditions of supply and demand existing in the financial market, as measured by a comparison of the volume of capital available for investment with the investment opportunities.

3. The prospect of inflationary or deflationary trends. Since bonds pay a fixed income throughout their duration, the real income derived therefrom will vary with the purchasing power of the dollar. Hence, bonds do not offer a suitable hedge against the possibility of inflation. If the investor anticipates a period of rising prices, he may prefer to purchase the common stock of a corporation, since the dividends and the market price of the stock will generally respond to a rising price structure. For this reason, the prospect of an inflationary period will tend to reduce the price which a bond can command.

4. In the particular case of municipal bonds, their tax-exemption fea-

ture is of intrinsic monetary value, thereby permitting these bonds to yield a low interest rate.   The value of the tax saving is of course dependent upon the tax rate to which the individual investor is subject and requires some appraisal of his future income.

5. In the case of convertible bonds, their yield is influenced by the value of the conversion feature.   This also requires an estimate of the future value of the stock, with all the attendant uncertainty.

In issuing bonds to the public, a corporation is assuming a long-term debt of a fixed nature, one that remains unaltered by changes in economic conditions.   In order to afford themselves a certain degree of flexibility, many corporations issue bonds containing the provision that at the option of the seller the bonds can be redeemed within a stated period prior to their maturity.   This clause is motivated by many considerations, such as the following:

1. If the bonds were issued when the credit standing of the corporation was weak, the corporation is burdened with paying a high interest rate to its bondholders for a protracted period of time, even though its financial standing may improve enormously subsequent to the sale of the bonds.

2. The corporation may wish to dispose of certain property that is pledged as security for a particular issue of bonds.

Bonds which confer upon the issuing corporation the option of retiring them prior to their maturity are termed "redeemable," or "callable," bonds.   As a partial compensation to the investor for the resulting inconvenience, redeemable bonds usually pay a premium if the redemption option is exercised.   The redemption value of the bond is generally expressed as a percentage of the par value.   Thus, if a $1,000 callable bond is to be redeemed at 103, the bondholder will receive $1,030 if the corporation calls the bond prior to. its maturity.

2-8. Savings and Loan Associations.   A savings and loan association is a mutually owned financial institution dedicated primarily to the investment of funds in urban mortgage loans.   Ownership of the institution is evidenced by "share accounts," which are usually of two types, "investment accounts" and "savings accounts."   The former are generally issued in units of $100, while the latter are issued for any sum.   The share-account holders elect the directors, who in turn elect the officers. Historically, the savings and loan association is an outgrowth of the old building and loan association, which was devoted to cooperative home financing.   The present-day institution, however, operates by accumulating funds from its owner-investors, which it then makes available to borrowers in the form of mortgage loans.

Each association is chartered by either the Federal government or the state in which it operates and is therefore subject to governmental regulations. Dividends are paid periodically to the investment-account holders but are often simply credited to the savings-account holders. Since the bulk of its funds are invested in mortgage loans, the association does not possess the same degree of liquidity enjoyed by savings banks. For this reason, the association generally requires thirty days' notice for the withdrawal of a large sum of money by an investor.

2-9. Indirect Forms of Investment. In the media of investment outlined above, the investment capital passed directly from the investor to the borrower of the capital. There are many forms of investment, however, in which the investment capital follows a somewhat circuitous path, passing through intermediate stages before it eventually reaches the borrower. For example, when an individual deposits money in a savings bank, the bank is the actual investment agent, utilizing the accumulated funds in the manner it deems fit, albeit within the limits prescribed by law.

Many indirect, or secondary, forms of investment also differ from the primary forms in the respect that they are not characterized basically by the interest-earning motive. For example, when an individual purchases a life insurance policy, his primary purpose is to assure his family of financial sustenance in the event of his demise. However, the insurance company invests the accumulated funds in various manners, with the policyholders obtaining a return on these investments in the form of a lower insurance rate or, in the case of a mutually owned company, in the form of periodic dividends. In recent years, life insurance companies have come to play an increasingly prominent role in the investment market.

CHAPTER 3

MONEY AS A FUNCTION OF TIME

3-1. Simple and Compound Interest. The interest-earning capacity of money dictates that any sum of money not required for immediate use should not be left idle but rather should be invested as soon as received. It therefore follows that the interest earned by a loan should theoretically be paid to the creditor on the date it becomes due to make it available to him for investment purposes. However, since the amount of the interest earned each period is small compared to the principal of the loan, it is frequently more convenient for both the creditor and the debtor to resort to the following arrangement: At the end of each interest period, the interest earned during that period is not paid directly but rather is added to the amount of the loan. In other words, the interest is converted to principal. This conversion occurs at the end of each interest period. Since interest is usually not paid for fractional parts of an interest period, the principal of a loan remains constant in the course of an interest period, and is augmented abruptly on the last day of the period through the conversion of interest.

As an illustration, assume that Smith borrows $3,000 from Jones for a period of 5 years, with interest payable at 6 per cent per annum. Interest earned is to be converted to principal at the end of each year. At maturity, the debt is to be discharged by means of a single payment. What is the amount of this payment?

At the expiration of the first year, the loan has earned interest of $180 (6 per cent of $3,000), and this interest is now added to the principal of the loan. The total principal at the commencement of the second year is therefore $3,180. At the end of this year, the interest earned is $190.80 (6 per cent of $3,180). This interest is also converted to principal, enhancing the total principal to $3,370.80. Table 3-1 records the growth of the principal of the loan.

The payment required at the end of the fifth year to discharge the debt

TABLE 3-1. PRINCIPAL OF LOAN

Year	Principal at beginning	Interest earned	Principal at end
1	$3,000.00	$180.00	$3,180.00
2	3,180.00	190.80	3,370.80
3	3,370.80	202.25	3,573.05
4	3,573.05	214.38	3,787.43
5	3,787.43	227.25	4,014.68

is therefore $4,014.68. The original principal itself has earned interest of $180 for 5 years, or a total of $900. The remaining interest of $114.68 was derived by the periodic conversion of this primary interest to principal.

The process of converting interest to principal is known as "compounding of interest." The interval between successive compoundings is known as the "period of compounding." We shall consider this period of compounding to be the actual interest period. Thus, if interest is compounded semiannually, then both the period of compounding and the interest period are 6 months.

Evidently, the more frequently interest is compounded, the more economical is the arrangement from the viewpoint of the lender. However, since the clerical work involved in calculating and recording the interest earnings increases in direct proportion to the number of compoundings, it becomes necessary to set some practical limit to the period of compounding. In the case of money deposited in a savings bank, the usual period of compounding is 3 months.

Interest rates are generally expressed as though they applied to an annual period, even when the period of compounding (and therefore the true interest period) is a fractional part of a year. The true interest rate for each interest period is then obtained by dividing the annual rate by the number of compoundings per year. The stipulated annual rate is referred to as the nominal interest rate. Thus, the expression "interest at 6 per cent per annum compounded quarterly" means that the nominal rate is 6 per cent, that the true interest period is 3 months, and that the true interest rate pertaining to this period is 1½ per cent. Where the statement of an interest rate omits any mention of the frequency of compounding, it is understood that interest is compounded annually.

In the case of many short-term loans, where the debt is to be discharged by means of a single payment at maturity, the interest earned is neither compounded nor paid to the creditor at the expiration of the interest period but is simply held in abeyance until the maturity of the loan. Thus, for each interest period, the interest is calculated on the basis of the original principal. Since the principal remains constant, the periodic interest earning remains constant. Interest earned in this manner is known as "simple interest." At maturity, the debt is discharged by a single payment comprising both the principal and the total interest accrued. Such an arrangement for the repayment of a loan is resorted to solely for the sake of simplicity in calculating the interest earned by the loan.

In the example above, where Smith borrows $3,000 from Jones for a period of 5 years with interest payable at 6 per cent, assume that simple interest is earned. Then

Interest earned annually, 6 per cent of $3,000 = $ 180

Total interest earned for 5 years   = $ 900

Principal of loan   =   3,000

Total due at maturity   = $3,900

In general, let

P_0 = principal of the loan

i = interest rate per period (simple interest)

n = number of interest periods constituting the life of the loan

P_n = maturity value of loan

Then
$$P_n = P_0(1 + ni) \tag{3-1}$$

Simple interest has no true economic significance, and we shall deal solely with compound interest in studying the variation of money with time.

3-2. Basic Assumptions. Our study of finance is predicated on the fundamental fact that money, when properly invested, possesses the capacity to enlarge itself with the passage of time. Having recognized this fact, we proceed now to the problem of determining mathematically the manner in which money varies as a function of time. However, it is evident at the outset that in attacking this problem we face certain serious obstacles. In the first place, economics is a highly complex and dynamic subject, one that lends itself to precise mathematical analysis only to a limited extent. So many factors are interwoven in any given situation that it is impossible to study each of them in isolation. There is no way of anticipating the economic changes that may result from natural or technological developments occurring in the future. Second, we even lack a reliable unit for measuring the value of a given sum of money, since the value of a dollar as determined by its purchasing power fluctuates constantly. It is therefore apparent that, in order to enable us to undertake a mathematical analysis of the variation of money with time, it becomes necessary to make certain simplifying assumptions. We assume that:

1. Money never remains idle, but is invested directly following its receipt in a manner to yield an anticipated rate of return.

2. Every investment materializes in exactly the manner anticipated. For example, we assume that every loan is repaid at maturity, that corporate stocks continually pay the dividend rate expected, that every business venture yields the anticipated rate of return, etc.

3. The economic conditions obtaining at a given instant of time will remain immutable. As a consequence of this assumption, it is accepted that both the price structure and the interest rate prevailing at this instant of time remain permanently unaltered.

This assumption of static conditions is, of course, contrary to economic fact. The prices of commodities and the interest rate obtainable for money loaned are in a constant state of flux, responding to the expansion or contraction of both supply and demand. Scientific advances produce an endless stream of new products and improvements of existing ones. Our general mode of living is subject to steady modification. Moreover, with regard to the interest rate, not only does it fluctuate with time, but at any given instant the interest rate varies according to the type of loan, depending upon such factors as the magnitude of the principal available and the degree of risk involved in the venture. Nevertheless, it is necessary for us to disregard these considerations and, after having selected an appropriate interest rate for use in a particular set of calculations, to assume that it will remain constant.

The fact that we are assuming completely static business conditions does not imply that we are blinding ourselves to the existence of the dynamic factors in our economy. The assumption is made simply as a first step in the study of finance; it enables us to arrive at certain results in a straightforward mathematical manner. Being cognizant of these dynamic factors, however, we realize that it may be necessary to modify our mathematical results in order to give proper weight to such factors as inflationary or deflationary trends and the effects produced by technological developments, population trends, etc.

3-3. The Variation of Money with Time. The interest-earning capacity of money is not a characteristic that money possesses per se; money can only earn interest if it is properly invested. In other words, the ability to earn interest is a latent, not an inherent, property of money. However, as a result of our first two assumptions, we may regard money as if it did in fact possess this capacity intrinsically. From our point of view, therefore, money is a self-generating commodity, enlarging itself at a definite rate as time elapses. Consequently, assuming that the rate of return is identical for each form of investment, we are generally not concerned with the precise manner in which money is put to productive use, whether it be used to extend a loan, to invest in a business venture, to purchase corporate stock, etc. When defined in relation to this concept of the self-generating characteristic of money, the terms "principal" and "sum of money" become synonymous, as do the terms "interest rate" and "investment rate."

Money and time are inextricably linked in a two-dimensional coordinate system. To illustrate the significance of this fact, we shall employ an analogy from the physical sciences. Let us assume that we have a given quantity of radium, a substance that is continuously disintegrating at a definite rate. At the expiration of 1,590 years, one-half of the original quantity of radium remains; at the expiration of the next 1,590 years, one-quarter of the original quantity remains; etc. Since the quantity of radium present is continuously varying, it is evidently meaningless to express the amount of radium present without also expressing the time at which this amount existed.

Similarly, since we consider money to be constantly expanding at a definite rate, a given sum of money is growing in quantity as time elapses. Therefore, it is likewise meaningless to express the quantity pertaining to a given sum of money without also expressing the time at which the money possessed, or will possess, this quantity. Every statement of quantity must be coupled with a statement of the time corresponding thereto. We cannot conceive of money dissociated from a consideration of time any more than we can locate a point on the earth's surface by knowing its latitude but not its longitude. We shall henceforth refer to the date at which a given sum of money has a specified value as its "valuation date," or "zero date."

3-4. Equivalent Money-Time Expressions. Assume that an individual received the sum of $1,000 on Jan. 1, 1970, and immediately deposited this in a fund earning interest at the rate of 5 per cent per annum compounded annually. During a 3-year period, this principal expanded in the following manner:

Year	Principal at beginning	Interest earned	Principal at end
1970	$1,000.00	$50.00	$1,050.00
1971	1,050.00	52.50	1,102.50
1972	1,102.50	55.13	1,157.63

We thus have the following money-time expressions:

Quantity of money	*Valuation date*
$1,000.00	Jan. 1, 1970
1,050.00	Jan. 1, 1971
1,102.50	Jan. 1, 1972
1,157.63	Jan. 1, 1973

If this individual, instead of receiving the sum of $1,000 at the beginning of 1970, had received the sum of $1,050 at the beginning of 1971 or the sum of $1,157.63 at the beginning of 1973, his financial status would have been identical as far as his capital at any subsequent date is concerned. Hence, the receipt of $1,000 on Jan. 1, 1970, is equivalent to the receipt of $1,157.63 on Jan. 1, 1973, based on an interest rate of 5 per cent.

Consider now the reverse case, in which an individual made a disbursement of $1,000 on Jan. 1, 1970. Had he retained this money for an additional 3 years and kept it invested at a 5 per cent rate, it would have expanded to the sum of $1,157.63. Hence, a disbursement of $1,000 made on Jan. 1, 1970, is equivalent to a disbursement of $1,157.63 made on Jan. 1, 1973.

Summarizing these two cases, then, we can say that the sum of $1,000, valuation date Jan. 1, 1970, is equivalent to $1,157.63, valuation date Jan. 1, 1973, or that the sums of $1,000 and $1,157.63 are equivalent to one another for a time interval of 3 years.

The table above, which lists four equivalent money-time expressions, can be extended indefinitely, yielding an infinite number of equivalent expressions. The equivalence of these expressions is, of course, based exclusively on an interest rate of 5 per cent; for any other interest rate, their equivalence vanishes.

3-5. The Fundamental Money-Time Equation. In the preceding section, we traced the growth of the principal in a fund from its initial value of \$1,000 on Jan. 1, 1970, to its final value of \$1,157.63 on Jan. 1, 1973. To state this in another manner, we have translated the sum of \$1,000, valuation date Jan. 1, 1970, to its equivalent value of \$1,157.63, valuation date Jan. 1, 1973, based on an interest rate of 5 per cent. We shall now derive an equation to express mathematically the variation of money with time, using a stipulated interest rate.

Consider a sum of money to be deposited in a fund earning interest at the rate of i per period, the date of deposit being the first day of an interest period. We shall let P denote the principal in the fund, and since the principal is a function of time, we shall use a subscript attached to P to denote the time at which the principal is being determined. In all cases, our unit of time is one interest period. The principal in the fund will be evaluated at the end of an interest period, immediately after the interest earned during that period has been converted to principal. Thus, P_5 denotes the principal in the fund at the end of the fifth interest period, immediately after the compounding of the interest earned during the fifth period. In general, let

P_0 = quantity of money deposited in the fund
i = interest rate
P_n = principal in fund at end of the nth period

The history of the principal in the fund is recorded below:

Period No. 1

$$\begin{aligned}
\text{Principal at beginning} &= P_0 \\
\text{Interest earned} &= P_0 i \\
\text{Principal at end} &= \overline{P_0 + P_0 i} \\
P_1 &= \overline{P_0(1 + i)}
\end{aligned}$$

Period No. 2

$$\text{Principal at beginning} = P_0(1 + i)$$
$$\text{Interest earned} = P_0(1 + i)i$$
$$\text{Principal at end} = \overline{P_0(1 + i) + P_0(1 + i)i}$$
$$= P_0(1 + i)(1 + i)$$
$$P_2 = \overline{P_0(1 + i)^2}$$

Period No. 3

$$\text{Principal at beginning} = P_0(1 + i)^2$$
$$\text{Interest earned} = P_0(1 + i)^2 i$$
$$\text{Principal at end} = \overline{P_0(1 + i)^2 + P_0(1 + i)^2 i}$$
$$= P_0(1 + i)^2(1 + i)$$
$$P_3 = \overline{P_0(1 + i)^3}$$

In general, therefore,

$$P_n = P_0(1 + i)^n \tag{3-2}$$

This equation can also be derived in a slightly modified manner, as follows: Let P_r denote the principal in the fund at the end of the rth period, where r is a positive integer less than n. Then the interest earned during the $(r + 1)$st period is $P_r i$, and the principal at the end of the $(r + 1)$st period is the sum of these two amounts. Hence,

$$P_{r+1} = P_r + P_r i = P_r(1 + i)$$

Thus, the principal at the end of one interest period equals the principal at the end of the preceding period multiplied by the factor $(1 + i)$. Consequently,

$$P_1 = P_0(1 + i)$$
$$P_2 = P_1(1 + i) = P_0(1 + i)^2$$
$$P_3 = P_2(1 + i) = P_0(1 + i)^3$$
$$\cdot \cdot \cdot \cdot \cdot \cdot \cdot \cdot \cdot \cdot \cdot \cdot \cdot \cdot \cdot \cdot \cdot$$
$$P_n = P_0(1 + i)^n \tag{3-2}$$

In Chap. 1, we define a geometrical series as one in which the ratio of each term to the preceding term is constant. It is thus seen that the successive end-of-period values of the principal constitute such a series.

Values of the expression $(1 + i)^n$ are presented in the tables of Appendix B for the interest rates most frequently encountered in practice. Since Eq. (3-2) involves the independent variable n as an exponent, it is designated an exponential equation.

Example 3-1. If $1,000 is invested at an interest rate of 4 per cent per annum, what is its value at the end of 6 years?

Solution

$$P_0 = \$1,000 \qquad i = 4 \text{ per cent} \qquad n = 6$$
$$P_n = P_0(1 + i)^n = 1,000(1.04)^6$$

By Table B-9, $(1.04)^6 = 1.26532$.

$$\therefore P_6 = 1,000(1.26532) = \$1,265.32$$

Example 3-2. If $500 is deposited in a fund earning interest at the rate of 6 per cent compounded quarterly, what is the principal at the end of 3 years?

Solution

$$P_0 = \$500 \qquad i = 6 \text{ per cent}/4 = 1\tfrac{1}{2} \text{ per cent}$$
$$n = 3 \text{ years} \times 4 \text{ periods per year}$$

or

$$n = 12$$
$$P_{12} = 500(1.015)^{12} = 500(1.19562)$$
$$= \$597.81$$

Many problems involving the application of Eq. (3-2) arise in which P_0 and P_n are known and it is necessary to calculate either i or n. These problems can be most expeditiously solved by rewriting this equation in logarithmic form and using the table of logarithms, Table A-1 of the Appendix:

$$P_n = P_0(1 + i)^n \qquad\qquad (3\text{-}2)$$
$$\therefore \log P_n = \log P_0 + n \log (1 + i) \qquad\qquad (3\text{-}2a)$$

or

$$\log \frac{P_n}{P_0} = n \log (1 + i) \qquad\qquad (3\text{-}2b)$$

Example 3-3. On Jan. 1, 1966, Smith borrowed $5,000 from Jones and discharged the debt on Dec. 31, 1970, with a payment of $6,100. If we assume interest to be compounded annually, what was the interest rate implicit in this loan?

Solution

$$P_0 = \$5,000 \qquad P_5 = \$6,100 \qquad n = 5$$

Substituting in Eq. (3-2a),

$$\log 6,100 = \log 5,000 + 5 \log (1 + i)$$
$$\therefore \log (1 + i) = \frac{\log 6,100 - \log 5,000}{5}$$
$$= \frac{3.7853 - 3.6990}{5} = 0.0173$$
$$\therefore 1 + i = 1.041$$
$$i = 0.041 = 4.1 \text{ per cent (to the nearest tenth of one per cent)}$$

Example 3-4. If the sum of \$1,000 is invested at $4\frac{1}{4}$ per cent compounded annually, what will be the principal at the end of 10 years?

Solution

$$P_0 = \$1,000 \qquad n = 10 \qquad i = 4.25 \text{ per cent}$$

Since this interest rate is not included in our table, we must calculate P_n by applying the logarithmic equation:

$$\log P_n = \log P_0 + n \log (1 + i) \tag{3-2a}$$
$$\log P_{10} = \log 1,000 + 10 \log (1.0425)$$
$$= 3 + 10(0.0181) = 3.1810$$
$$P_{10} = \$1,517$$

In deriving Eq. (3-2), we defined P_n as the principal to which a given quantity of money P_0 will expand if invested in a fund at interest rate i. However, since we regard the interest-earning capacity as an inherent characteristic of money, we can extend our definition of P_n to say that it is simply the value of P_0 after n interest periods have elapsed, or that P_n is equivalent to P_0 for a time interval of n periods, based on an interest rate i. Thus, Eq. (3-2) is the basic money-time equation.

In many instances, we are primarily concerned not with the final value of the principal but rather with the amount of interest earned by the original sum of money. We will therefore derive an equation for this quantity. Let

I_n = interest earned by \$1 in n periods, based on an interest rate i

Now, the interest earned by a sum P_0 in n periods is

$$P_n - P_0 = P_0(1 + i)^n - P_0 = P_0[(1 + i)^n - 1]$$

Setting P_0 equal to \$1, we obtain

$$I_n = (1 + i)^n - 1 \tag{3-3}$$

3-6. Calculation of Discount. When we are given the value of a sum of money at a specified valuation or zero date, Eq. (3-2) enables us to translate that sum to its equivalent value at any valuation date in the future. It is often necessary, however, to reverse the process, that is, to translate a given sum of money to its equivalent value at a prior valuation date. To illustrate, assume that an individual is to receive the sum of \$5,000 three years from the present date. Should he require money for his immediate needs, he can obtain money for his claim by assigning this claim to another party. This party will pay the individual a sum of money whose value at the present date is equivalent to $5,000 three years hence. The two parties must agree, of course, on the rate to be used in calculating this equivalent value. This process of converting a claim on a future sum of money to a sum of money in the present is known as "discounting," and the difference between the two sums is termed the "discount."

Adhering to our previous notation, we shall let P_0 denote the value of a given sum of money at a specified valuation or zero date and let P_{-n} denote the value of P_0 at a valuation date n periods prior to that of P_0. Assume that the sum of money P_{-n} is deposited in a fund at interest rate i per period; by definition, its value at the end of n periods will be P_0. Applying Eq. (3-2) but substituting P_{-n} for P_0, and P_0 for P_n, we obtain

$$P_0 = P_{-n}(1 + i)^n$$

or

$$P_{-n} = \frac{P_0}{(1 + i)^n}$$

$$= P_0(1 + i)^{-n} \tag{3-4}$$

(The significance of negative exponents is explained in Chap. 1.)

If we compare this result with Eq. (3-2), it is seen that the latter can be extended to include Eq. (3-4) simply by assigning negative values to n. In summary, therefore, if we are given a quantity of money P_0 at a given zero date, its equivalent value P_n at any other date can be determined by substituting in the equation

$$P_n = P_0(1 + i)^n \tag{3-2}$$

If we wish to project the given sum P_0 into the future, we assign a positive value to n; if we wish to revert the sum P_0 to a date in the past, we assign a negative value to n.

Although Eq. (3-4) is theoretically superfluous, we shall nevertheless retain it as a distinct equation for problems involving the discount value of money. The tables of Appendix B present values of the expression $(1 + i)^{-n}$ for the interest rates most frequently encountered.

Example 3-5. An individual possesses two promissory notes, each having a maturity value of $1,000. The first note is due 2 years hence, and the second is due 3 years hence. At an interest rate of 6 per cent, what proceeds will he obtain by discounting the notes at the present date?

Solution

$$P_{-n} = P_0(1 + i)^{-n}$$

Value of first note:

$$P_{-2} = 1,000(1.06)^{-2} = 1,000(0.89000) = \$890.00$$

Value of second note:

$$P_{-3} = 1,000(1.06)^{-3} = 1,000(0.83962) = \$839.62$$
$$\text{Total discount value} = \$890.00 + \$839.62 = \$1,729.62$$

In commercial practice, discount is calculated in a manner different from that presented above, or, to view the situation in another way, the term "discount rate" is used in a sense that is distinct from "interest rate." To illustrate, assume that a note whose maturity value is $500 is to be discounted 1 year before maturity at an interest rate of 6 per cent. The discount value is

$$P_{-1} = 500(1.06)^{-1} = 500(0.94340) = \$471.70$$

When based on an interest rate of 6 per cent, the two sums of $471.70 and $500 are equivalent to one another for a time interval of 1 year. If the smaller sum is invested at 6 per cent, it earns an interest for 1 year of $28.30, thereby amounting to $500 at the end of the year. Thus the interest rate is always applied to the smaller, or prior, value to obtain the larger, or future, value. On the other hand, when the term "discount rate" is used commercially, this rate is applied to the future value to obtain the prior value. Thus, if a note of $500 maturity value is discounted at a bank 1 year in advance at a discount rate of 6 per cent, the calculation is

Maturity value	= $500
Discount, 6 per cent of $500 =	30
Discount value	= $470

Evidently, if the sum of $470 is to increase to $500 at the end of 1 year, it must be invested at a rate of 6.38 per cent. Thus, an interest rate of 6 per cent and a discount rate of 6 per cent produce dissimilar results; the two terms are therefore not synonymous.

For our present study of finance, it would not only be inconvenient to use two distinct rates, one for the calculation of interest and one for the calculation of discount, but it is theoretically incorrect to do so. Having selected a specific interest rate, we must use that rate consistently in our calculations, regardless of whether we are progressing or retrograding in time.

Thus, if the sum of $3,000 is invested at an interest rate of 5 per cent, its value at the end of 10 years is $4,887. For a time interval of 10 years, therefore, the sums of $3,000 and $4,887 are equivalent to one another, based on an interest rate of 5 per cent. Their difference of $1,887 can be regarded as either the interest earned by $3,000 in 10 years or as the discount applicable to $4,887 for 10 years. In contrast to commercial practice, we are using in this text only one interest rate to determine equivalent money values. Let

D_n = discount applicable to $1 for n periods at interest rate i

Now, the discount applicable to the sum P_0 for n periods is

$$P_0 - P_{-n} = P_0 - P_0(1 + i)^{-n} = P_0[1 - (1 + i)^{-n}]$$

Setting P_0 equal to $1, we obtain

$$D_n = 1 - (1 + i)^{-n} \tag{3-5}$$

3-7. The Rate of Growth of Principal. Although the expression "interest rate" is universally employed to designate the ratio of interest earned per period to principal, the use of this term can be somewhat misleading; it would be more correct to use the expression "interest ratio." The interest rate does not represent the rate at which a given principal increases in value with the passage of time. Evidently, since we are using one interest period as our unit of time, the time rate at which a given principal expands in the course of a particular period is numerically equal to the interest earned during that period. Hence,

Time rate of growth of principal = interest earned per period
= principal at beginning of period $\times i$

where i is the interest rate. The beginning-of-period principal is a variable quantity, while i remains constant. The rate of growth of principal is therefore directly proportional to the principal at the commencement of the period, and the interest rate i is the constant of proportionality.

3-8. Graphical Representation. There are many advantages inherent in the process of representing variable quantities in a graphical manner, and we shall employ this device extensively in this text. The variation of the value of money with the passage of time can be charted on a money-time (or principal-time) graph, in which time is the abscissa and the value of a given sum of money is the ordinate.

In Fig. 3-1, line OQ_0, on the YY axis, represents a sum of money P_0 at a given valuation, or zero, date. The equivalent values of P_0 at subsequent valuation dates are represented by the ordinates A_1Q_1, A_2Q_2, etc., while the equivalent values of P_0 at prior valuation dates are represented by ordinates $A_{-1}Q_{-1}$, $A_{-2}Q_{-2}$, etc. In order to accentuate the general features of the principal-time diagram, we have selected a very high

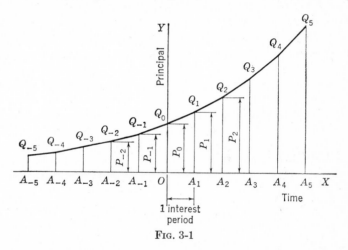

Fig. 3-1

interest rate, namely, 25 per cent. Assume for convenience that interest is paid for fractional parts of an interest period. On this basis, although the interest-earning principal remains constant in the course of an interest period, the value of the given sum of money is continuously increasing, being the sum of the principal at the beginning of the period and the interest accrued to that particular date.

During the first period, the money expands uniformly along line Q_0Q_1, from an initial value of P_0 to a final value of P_1. Hence, the slope of line Q_0Q_1 measures the rate of expansion. During the second period, the money expands uniformly along line Q_1Q_2, from its initial value of P_1 to its final value of P_2. Line Q_1Q_2 has a greater inclination than line Q_0Q_1. It is evident, therefore, that the principal-time graph is composed of a series of straight lines, with a distinct line applying to each period.

Although Fig. 3-1 represents the true history of the variation of the value of money, it is too cumbersome to construct when the number of interest periods is large. For convenience, therefore, we shall replace this graph by a smooth curve connecting the end-of-period values, represented by the various Q points. This is tantamount to treating the equation

$$P_n = P_0(1 + i)^n \qquad (3\text{-}2)$$

as if it applied to fractional as well as integral values of n. The curve obtained by connecting the successive end-of-period values is shown in

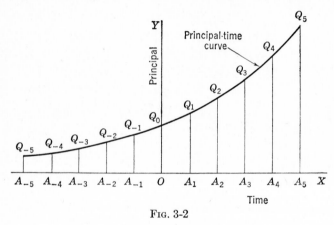

Fig. 3-2; we shall henceforth refer to this as the principal-time curve. It is designated an exponential curve, since it is the graph of an exponential equation.

There is a distinct principal-time curve corresponding to every interest rate. Obviously, the greater the interest rate, the steeper the principal-time curve. As a matter of fact, since the principal is a function of both time and interest rate, the complete graphical representation of principal really requires a three-dimensional coordinate system, in which time is measured along one axis, the interest rate along the second axis, and the principal along the third axis. For our present study, however, the two-dimensional principal-time curve will satisfy our requirements.

In Fig. 3-3, the principal-time curve has been based on a given principal P_0 at zero date and a stipulated interest rate. (In constructing this diagram we have again selected a high interest rate in order to bring into sharp focus the essential features of the graph.) A line $X'X'$ is drawn at point Q_0 parallel to axis XX. If we select a point Q_r on the curve whose abscissa is r interest periods, then the ordinate A_rQ_r represents the value of the original principal P_0 at the expiration of r interest periods, while the line T_rQ_r represents the interest earned by P_0 in r periods. Likewise, if we select a point Q_s on the curve whose abscissa is $-s$ periods, then the ordinate A_sQ_s represents the value of P_0 at a date s periods prior to that of P_0, while the line T_sQ_s represents the discount applicable to P_0 for s periods. This quantity T_sQ_s can also be regarded as the interest earned by P_0 in $-s$ periods.

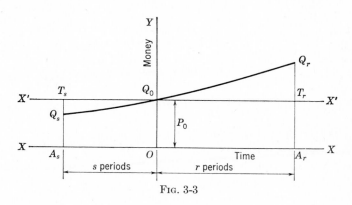

Fig. 3-3

Hence, with regard to curve Q_sQ_r, if we select axis XX as datum line, the ordinates of this curve represent the principal at their respective dates; but if we select axis $X'X'$ as datum line, then the ordinates of the curve represent interest earned by the sum P_0 at their respective dates. Thus, by a proper choice of datum line, the same curve can serve as either a principal-time curve or an interest-time curve. The importance of an interest-time curve will become apparent in our study of sinking funds.

Thus far, we have drawn all principal-time diagrams with the ordinate representing the given principal P_0 lying on the YY axis, or, to express this in another manner, with the zero date of P_0 coinciding with the zero date of the diagram. However, it is not always possible to adhere to this practice. For example, we may wish to represent on the same diagram two distinct sums of money having different zero dates. Here we are not dealing with time as an absolute quantity but rather as a relative quantity. We are solely concerned with the interval of time separating two given dates. Thus, in the equation

$$P_n = P_0(1 + i)^n \tag{3-2}$$

n is simply the difference in time between the valuation dates corresponding to P_n and P_0. It is therefore evident that the vertical ordinate representing P_0 can be placed at any location on the diagram without altering the nature of the principal-time curve, as long as the value of n is measured from this ordinate.

3-9. Method of Chronological Sequence and Method of Separate Evaluation. Assume that the following transactions occurred in a fund whose interest rate is 8 per cent per annum:

1. An initial deposit of $1,000 was made Jan. 1, 1960.

2. A withdrawal of $600 was made Jan. 1, 1964.

3. A deposit of $800 was made Jan. 1, 1969.

If the account was closed on Dec. 31, 1970, what was the final principal?

In solving this problem, we shall first assume that we are interested in knowing, not merely the final principal, but the principal at all intermediate dates as well. The history of the principal in the fund is depicted in Fig. 3-4.

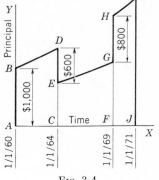

Fig. 3-4

There are four significant dates involved, and the principal at these dates can be calculated in the following manner:

$$AB = \$1,000$$
$$CD = 1,000(1.08)^4 = 1,000(1.36049) \qquad = \$1,360.49$$
$$CE = CD - ED = \$1,360.49 - \$600 \qquad = \$\ \ 760.49$$
$$FG = 760.49(1.08)^5 = 760.49(1.46933) \qquad = \$1,117.41$$
$$FH = FG + GH = \$1,117.41 + \$800 \qquad = \$1,917.41$$
$$JK = 1,917.41(1.08)^2 = 1,917.41(1.1664) = \$2,236.47$$
$$\therefore \text{ Principal on } 12/31/70 = \$2,236.47$$

The principal at any nonsignificant date can be determined by applying Eq. (3-2), using for P_0 the principal at the previous significant date.

We have thus arrived at the final principal by a step-by-step procedure, taking each transaction in its proper sequence and combining it with the transactions and interest earnings that have preceded it. We shall refer to this method of calculating principal as the method of "chronological sequence."

Assume next that we are interested solely in calculating the final principal, without regard to its value at intermediate dates. For this purpose, the method of solution followed above is unduly laborious. An alternate method of solution can be developed in the following manner: If we proceed with our solution in the same manner as previously, but without performing the actual arithmetical calculations, we obtain the following results:

$$AB = \$1,000$$
$$CD = 1,000(1.08)^4$$
$$CE = CD - ED = 1,000(1.08)^4 - 600$$
$$FG = [1,000(1.08)^4 - 600](1.08)^5 \text{ or } FG = 1,000(1.08)^9 - 600(1.08)^5$$
$$FH = FG + GH \text{ or } FH = 1,000(1.08)^9 - 600(1.08)^5 + 800$$
$$JK = [1,000(1.08)^9 - 600(1.08)^5 + 800](1.08)^2 \text{ or } JK = 1,000(1.08)^{11}$$
$$- 600(1.08)^7 + 800(1.08)^2$$

Before completing the arithmetical solution, we shall analyze this equation for the value of JK, for it discloses the effect produced on the final principal by the three operations performed upon the fund, namely, the two deposits and the one withdrawal. It is seen that each term in this equation is the value of the respective deposit or withdrawal as of Dec. 31, 1970. Thus, $1,000(1.08)^{11}$ is the value of the original deposit of \$1,000 when translated to the new valuation date. This would be the principal in the fund if no other transactions had occurred. The same applies to the term $800(1.08)^2$. Similarly, the term $600(1.08)^7$ represents the value at the end of 1970 of the sum withdrawn in 1964. Obviously, a withdrawal of \$600 at the beginning of 1964 has the same effect as a withdrawal of $600(1.08)^7$ at the end of 1970, as far as the principal at the latter date or any subsequent date is concerned.

Completing the arithmetical solution, we obtain

$$
\begin{aligned}
1,000(1.08)^{11} = 1,000(2.33164) &= \$2,331.64 \\
800(1.08)^2 = 800(1.16640) &= \underline{933.12} \\
\text{Total} &= \$3,264.76 \\
600(1.08)^7 = 600(1.71382) &= \underline{1,028.29} \\
\text{Principal on } 12/31/70 &= \underline{\underline{\$2,236.47}}
\end{aligned}
$$

It is evident from the above discussion that when a fund has been subjected to a series of deposits and withdrawals and we are solely concerned with calculating the principal at one particular date, it is unnecessary to trace in exact detail the complete history of the fund, as was done in the first method of solution. In lieu of that, we can isolate each deposit and withdrawal, translate the respective sums of money involved to their equivalent values at the date the principal is being determined, and then obtain the algebraic sum of these equivalent values. We shall refer to this method of calculating the principal as the method of "separate evaluation."

The graphical representation of the principal in the fund, based on the method of separate evaluation, is presented in Fig. 3-5. Each sum of money is treated in isolation from the others, and a separate principal-time curve is drawn for each sum.

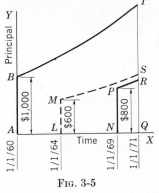

FIG. 3-5

The curve representing the value of the withdrawal is shown as a broken line to distinguish it from the deposits.

In the solution of numerical problems, we shall generally apply the method of separate evaluation since, of the two methods available, this entails the lesser amount of arithmetical work. However, there are many cases in which we are interested in determining not simply the value of the principal at a specific date but also the manner in which the principal varies in the course of time. Thus, a mortgagor is concerned not only with the amount of his outstanding debt at a particular instant of time but also with the rate at which his periodic payments are liquidating the debt. If these problems are represented graphically by the method of chronological sequence, then the rate of variation of principal is readily discernible from the graph, for only one line is used to represent the principal. For this reason, we shall frequently apply the method of chronological sequence in constructing the principal-time graphs.

3-10. Equivalent Groups of Money Values. If we are given the value of a particular sum of money at a specific date, we have seen that it is possible to translate this value to an equivalent value at any other date, based on a stipulated interest rate. This rule applies not only to individual sums of money but to groups of money values as well. The problem solved in the preceding section is an illustration of this fact. In calculating the final principal we have, in effect, translated the three money values involved—the deposits of $1,000 and $800 and the withdrawal of $600—to an equivalent single money value of $2,236.47 on Dec. 31, 1970. There are also many cases where it is necessary to translate a given group of money values to a second group equivalent to the first at another date.

Example 3-6. Smith owes Jones the sum of $3,000, due Dec. 31, 1982, and $2,000, due Dec. 31, 1984. By mutual consent, the terms of payment are altered to allow Smith to discharge the debt by making a payment of $4,000 on Dec. 31, 1985, and a payment for the balance on Dec. 31, 1986, with interest at 5 per cent. What will be the amount of this final payment?

Solution. The given data are tabulated below. The final payment, whose value is to be calculated, is denoted by X.

Payments—Original plan	Payments—Revised plan
$3,000 on 12/31/82	$4,000 on 12/31/85
$2,000 on 12/31/84	X on 12/31/86

Since the two groups of payments are equivalent to one another, based on an interest rate of 5 per cent, then the total value of one group must equal the total value of the other group at every instant of time. To determine X, it is simply necessary to select some standard date for evaluating all sums of money involved. For convenience, we shall select Dec. 31, 1986, as the valuation date. We then have

$$4,000(1.05) + X = 3,000(1.05)^4 + 2,000(1.05)^2$$
$$4,000(1.05) + X = 3,000(1.21551) + 2,000(1.10250)$$
$$4,200 \quad\quad + X = 3,646.53 + 2,205$$
$$X = \$1,651.53$$

This result, of course, is independent of the valuation date selected. Thus, assume we had selected Dec. 31, 1982. Our solution would then be

$$4,000(1.05)^{-3} + X(1.05)^{-4} = 3,000 + 2,000(1.05)^{-2}$$
$$4,000(0.86384) + X(1.05)^{-4} = 3,000 + 2,000(0.90703)$$
$$3,455.36 \quad\quad + X(1.05)^{-4} = 3,000 + 1,814.06$$
$$X(1.05)^{-4} = 1,358.70$$
$$X = 1,358.70(1.05)^4 = 1,358.70(1.21551)$$
$$= \$1,651.51$$

That this payment is the correct amount required to liquidate the debt can be verified by applying the method of chronological sequence to obtain the following calculations:

$$
\begin{aligned}
\text{Value of 1st debt on } 12/31/84 = 3,000(1.05)^2 &= \$3,307.50 \\
\text{Value of 2d debt on } 12/31/84 &= \underline{2,000.00} \\
\text{Total principal of loan on } 12/31/84 &= \$5,307.50 \\
\text{Interest earned in 1985} &= \underline{265.38} \\
\text{Principal of loan on } 12/31/85 &= \$5,572.88 \\
\text{Payment on } 12/31/85 &= \underline{4,000.00} \\
\text{Principal of loan on } 1/1/86 &= \$1,572.88 \\
\text{Interest earned in 1986} &= \underline{78.64} \\
\text{Principal of loan on } 12/31/86 &= \$1,651.52 \\
\text{Payment required on } 12/31/86 &= \underline{1,651.52}
\end{aligned}
$$

It is to be emphasized that these two groups of payments are equivalent to one another only for an interest rate of 5 per cent. Implicit in our reasoning is the assumption that the creditor can reinvest each sum of money he receives in a manner that continues to yield 5 per cent. Should there occur any variation of the interest rate, then the equivalence of the two groups of payments is destroyed. For example, assume that the creditor is able to reinvest his capital at a rate of only 4 per cent. For this situation, the revised terms of payment are more advantageous to him since, by deferring the collection of the money due him, he enables his capital to earn the higher rate of interest for a longer period of time.

3-11. Calculation of an Investment Rate. In the preceding problem, the replacement of one group of money values with an equivalent one was accomplished with a known interest rate. Many problems arise in practice, however, in which the equivalent groups of money values are known, and it is necessary to determine the interest rate on which their equivalence is predicated. Problems of this nature can only be solved by a trial-and-error method, and if the exact interest rate is not one of those listed in the interest tables, it can be approximated by means of straight-line interpolation.

Example 3-7. Smith owed Jones the sum of $1,000, due Dec. 31, 1970, and $4,000, due Dec. 31, 1972. Because Smith was unable to meet these obligations as they became due, the debt was discharged by means of a payment of $2,000 on Dec. 31, 1973, and a second payment of $3,850 on Dec. 31, 1974. What annual interest rate was intrinsic in these payments?

Solution. Let i denote the interest rate. If the expression $(1 + i)^n$ is expanded by the binomial theorem and all terms beyond the second are discarded, we obtain

$$(1 + i)^n = \text{approximately } 1 + ni$$

We shall apply this approximate value to obtain a first approximation of the value of i. This procedure is tantamount to basing our first calculation on the use of simple rather than compound interest. This approximation understates the value of $(1 + i)^n$, and the degree of error varies in proportion to n.

Since the two groups of payments in this problem are equivalent, the value of one group equals the value of the other at any valuation date that we select. Equating the two groups at the end of 1974 on the basis of simple interest, we obtain

$$[1,000 + 4i(1,000)] + [4,000 + 2i(4,000)] = [2,000 + i(2,000)] + 3,850$$

Solving for i,

$$i = 8.5 \text{ per cent} \qquad \text{as a first approximation}$$

Now, the use of this approximation in our calculations has produced an understatement of the value of each sum of money except the last. Moreover, since the amount of error increases as n increases, it is evident that we have reduced the importance of the money values having early valuation dates and inflated the importance of those having late valuation dates. The true rate, therefore, will be less than $8\frac{1}{2}$ per cent. Assume an 8 per cent rate. If this were the actual rate, then the payment X required at the end of 1974 is determined as follows:

$$1,000(1.08)^4 + 4,000(1.08)^2 = 2,000(1.08) + X$$
$$X = \$3,866.09$$

Since the actual payment was $3,850, the true interest rate is less than 8 per cent. Try a $7\frac{1}{2}$ per cent rate.

$$1,000(1.075)^4 + 4,000(1.075)^2 = 2,000(1.075) + X$$
$$X = \$3,807.99$$

The interest rate therefore lies between 8 per cent and $7\frac{1}{2}$ per cent. Applying straight-line interpolation, we obtain

$$i = 7.86 \text{ per cent}$$

The problem of calculating an interest rate is often encountered where an investment is made that produces certain known returns at future dates and we wish to determine the rate of return earned by the investment.

Example 3-8. (P.E. examination problem.) A lot was purchased in January, 1960, for $2,000. Taxes and assessments were charged at the end of each year as follows: $30 in 1960, $30 in 1961, $200 in 1962, $30 in 1963, $30 in 1964, and $30 in 1965. The owner paid the charges for 1960 and 1961, but not for subsequent years. At the end of 1965, the lot was sold for $3,000, the seller paying the back charges at 7 per cent interest compounded annually and also paying a commission of 5 per cent to an agent. What rate of return was realized on the investment?

Solution. The payment made by the investor at the end of 1965 for the back charges consisted of the following:

For 1962, $200(1.07)^3$	= $	245.01
For 1963, $30(1.07)^2$	=	34.35
For 1964, $30(1.07)$	=	32.10
Total back charges	= $	311.46
Taxes for 1965	=	30.00
Commission, 5 per cent of $3,000	=	150.00
Total payment at end of 1965	= $	491.46
Selling price	=	3,000.00
Net receipt at end of 1965	=	$2,508.54

We thus obtain the following two equivalent groups of money values:

Group 1—Disbursements	*Group 2—Receipts*
$2,000 on 1/1/60	$2,508.54 on 12/31/65
$30 on 12/31/60	
$30 on 12/31/61	

Let i denote the investment rate. Using simple interest to obtain a first approximation and selecting Dec. 31, 1965, as our valuation date, we obtain

$$[2,000 + 6i(2,000)] + [30 + 5i(30)] + [30 + 4i(30)] = \$2,508.54$$

Solving for i,

$$i = 3.66 \text{ per cent} \qquad \text{as a first approximation}$$

Since our approximate method of solution has given insufficient weight to the disbursements, the true investment rate is less than this. Try a $3\frac{1}{2}$ per cent rate. The income X required to yield this rate is found as follows:

$$2,000(1.035)^6 + 30(1.035)^5 + 30(1.035)^4 = X$$
$$X = \$2,528.58$$

Since the actual income was \$2,508.54, the investment rate is less than $3\frac{1}{2}$ per cent. Try a 3 per cent rate.

$$2,000(1.03)^6 + 30(1.03)^5 + 30(1.03)^4 = X$$
$$X = \$2,456.65$$

Applying straight-line interpolation, we obtain

$$i = 3.36 \text{ per cent}$$

3-12. Effect of Taxes on Investment Rate. Risk is inherent in every investment, and therefore, a proposed investment should be undertaken only if it offers a certain minimum rate of return. Since most income earned in the United States is subject to some form of taxation, the minimum acceptable rate of return should be based on the income that remains after payment of taxes.

Although a detailed discussion of current tax laws is neither possible nor feasible in this text, the following points must be emphasized:

1. The tax laws distinguish between regular income, which is income that accrues from ordinary business operations, and long-term capital gains, which is the profit realized through the sale of capital assets that have been held for a certain minimum length of time. The two forms of income are subject to different tax rates.

2. Since income tax is a graduated tax, the rate at which the income from the proposed investment will be taxed is a function of the investor's present level of income. This rate should not be confused with the average tax rate that applies to the investor's present income.

To illustrate the second point, assume that the following tax rates apply to an individual's combined Federal and state income taxes:

Taxable income	*Tax*
\$20,000 to \$22,000	\$5,800 plus 40 per cent of amount above \$20,000
\$32,000 to \$38,000	\$11,100 plus 53 per cent of amount above \$32,000

Now assume that the proposed investment will yield an annual income of \$1,000. If the investor's current taxable income is \$20,000, his average tax rate is 29 per cent. However, the incremental income of \$1,000 will be taxed at the rate of 40 per cent, and the residual income will be \$600. If the investor's current taxable income is \$32,000, his average tax rate is 34.7 per cent. However, the incremental income will be taxed at the rate of 53 per cent, and the residual income will be \$470.

We shall now develop the relationship between the before-tax and after-tax investment rate. Let

$$C = \text{capital invested}$$
$$I = \text{annual income from investment}$$
$$t = \text{rate at which this income is taxed}$$
$$i_b \text{ and } i_a = \text{investment rate as calculated before and after payment of tax, respectively}$$

Original income $= I$ $\qquad\qquad i_b = \dfrac{I}{C}$

Residual income $= I(1-t)$ $\qquad i_a = \dfrac{I(1-t)}{C}$

Then $\qquad\qquad\qquad\qquad i_a = i_b(1-t)$ $\qquad\qquad\qquad$ (3-6)

Example 3-9. An investor wishes to earn 7 per cent on his capital after payment of taxes. If the income from an available investment will be taxed at an average rate of 42 per cent, what minimum rate of return, before payment of taxes, must the investment offer to be justified?

Solution

$$i_a = 0.07 \qquad t = 0.42$$
$$i_b = \frac{i_a}{1-t} = \frac{0.07}{0.58} = 0.121 = 12.1 \text{ per cent}$$

Owing to the many variables and uncertainties that are present, it is often difficult to estimate the rate at which the income from an investment will be taxed. Moreover, if the venture includes many individuals, this tax rate varies among the individuals. In the subsequent material, it is to be understood in the absence of any explicit statement to the contrary that the stipulated investment rate is the rate as calculated before payment of taxes.

3-13. Effective Interest Rate. Consider the sum of $1,000 to be deposited in a fund earning interest at the rate of 8 per cent per annum compounded quarterly. As previously stated, 8 per cent is merely a nominal interest rate; the true interest rate is 2 per cent per quarterly period. At the end of 6 months, or two interest periods, this sum has expanded to

$$P_2 = 1,000(1.02)^2 = \$1,040.40$$

If this fund, instead of earning 2 per cent interest per quarterly period, were earning interest at the rate of 4.04 per cent per semiannual period, then the principal at the end of any 6-month period would be the same. Hence, assuming that the principal is not withdrawn from the fund except at the end of a 6-month period, we may regard the two interest rates as being equivalent to one another.

At the expiration of 1 year, or four interest periods, the original sum of $1,000 has expanded to

$$P_4 = 1,000(1.02)^4 = \$1,082.43$$

Similarly, if this fund earned interest at the rate of 8.243 per cent compounded annually, the principal at the end of each year would be the same. This interest rate, then, is also equivalent to the given rate.

We have thus demonstrated the equivalence of these three interest rates:

2 per cent per quarterly period
4.04 per cent per semiannual period
8.243 per cent per annual period

If a given interest rate applies to a period less than 1 year, then its equivalent rate for an annual period is referred to as its "effective rate." Thus, the effective rate corresponding to a rate of 2 per cent per quarterly period is 8.243 per cent. The effective rate is numerically equal to the interest earned by a principal of $1 for 1 year. In general, let

j = nominal interest rate
r = effective interest rate
m = number of compoundings per year

Then
$$r = \left(1 + \frac{j}{m}\right)^m - 1 \tag{3-7}$$

Assume we wish to make a ratio comparison of the following two interest rates:

1st rate, 6 per cent per annum compounded monthly
$$= \tfrac{1}{2} \text{ per cent per month}$$
2nd rate, 4 per cent per annum compounded semiannually
$$= 2 \text{ per cent per semiannual period}$$

Since the two rates apply to unequal periods of time, a direct comparison is not possible. However, we can establish a basis of comparison by determining the effective rate corresponding to each:

Effective rate for 6 per cent compounded monthly
$$= r_1 = (1.005)^{12} - 1 = 6.168 \text{ per cent}$$
Effective rate for 4 per cent compounded semiannually
$$= r_2 = (1.02)^2 - 1 = 4.04 \text{ per cent}$$

$$\therefore \frac{r_1}{r_2} = \frac{6.168 \text{ per cent}}{4.04 \text{ per cent}} = 1.527$$

The calculation of effective interest rates is therefore highly useful for comparative purposes.

3-14. Adjustment of Monetary Values for Inflation. Although our study of finance rests on the assumption of static economic conditions, the anticipated effects of inflation or deflation can readily be included in our calculations where necessary. The notational system is as follows:

C_0 = present cost of given commodity
C_r = cost of given commodity r years hence
i_r = rate of inflation during rth year

The annual rate of inflation with respect to the given commodity is the ratio of the increase in cost during the year to the cost at the beginning of the year. Then

$$i_r = \frac{C_r - C_{r-1}}{C_{r-1}} \tag{3-8}$$

Assume that the annual rate of inflation remains constant for n years, and let i denote this rate. Proceeding as in Art. 3-5, we obtain the following:

$$C_n = C_0(1 + i)^n \tag{3-9}$$

If deflation is anticipated, i is assigned a negative value.

Example 3-10. A machine has been purchased and installed at a total cost of $18,000. The machine will be retired at the end of 5 years, at which time it is expected to have a scrap value of $2,000 based on current prices. The machine will then be replaced with an exact duplicate. The firm plans to establish a reserve fund to accumulate the capital needed to replace the machine. If an average annual rate of inflation of 3 per cent is anticipated, how much capital must be accumulated?

Solution. Both the cost of the machine and the scrap value will appreciate at the assumed average rate. The net cost of replacing the machine with a duplicate 5 years hence will be

$$(18,000 - 2,000)(1.03)^5 = 16,000(1.15927) = \$18,548$$

Example 3-11. A product has a current selling price of $325. If its selling price is expected to decline at the rate of 10 per cent per annum because of obsolescence, what will be its selling price 4 years hence?

Solution

$$\text{Anticipated selling price} = 325(1 - 0.10)^4 = 325(0.90)^4$$
$$= \$213.20$$

The solution can be verified readily by constructing a schedule showing the selling price at the end of each year during the 4-year period.

PROBLEMS

3-1. If the sum of $1,500 is deposited in an account earning interest at the rate of 5 per cent compounded quarterly, what will be the principal at the end of 7 years?
Ans. $2,124

3-2. An individual possesses a promissory note, due 2 years hence, whose maturity value is $3,200. What is the discount value of this note, based on an interest rate of 7 per cent?
Ans. $2,795

3-3. On July 1, 1962, a bank account had a balance of $8,000. A deposit of $1,000 was made on Jan. 1, 1963, and a deposit of $750 on Jan. 1, 1964. On Apr. 1, 1965, the sum of $1,200 was withdrawn. What was the balance in the account on Jan. 1, 1968, if interest is earned at the rate of 4 per cent compounded quarterly?
Ans. $10,719

3-4. A manufacturing firm contemplates retiring an existing machine at the end of 1982. The new machine to replace the existing one will have an estimated cost of $10,000. This expense will be partially defrayed by sale of the old machine as scrap for $750. To accumulate the balance of the required capital, the firm will deposit the following sums in an account earning interest at 5 per cent compounded quarterly:

$1,500 at the end of 1979
$1,500 at the end of 1980
$2,000 at the end of 1981

What cash disbursement will be necessary at the end of 1982 to purchase the new machine?
Ans. $3,750

3-5. A father wishes to have $5,000 available at his son's eighteenth birthday. What sum should be set aside at the son's fifth birthday if it will earn interest at the rate of 3 per cent compounded semiannually?   *Ans.* $3,395

3-6. In the preceding problem, if the father will set aside equal sums of money at the son's fifth, sixth, and seventh birthdays, what should each sum be?

3-7. Brown owes Smith the following sums:

$1,000 due 2 years hence
$1,500 due 3 years hence
$1,800 due 4 years hence

Having received an inheritance, he has decided to liquidate the debts at the present date. If the two parties agree on a 5 per cent interest rate, what sum must Brown pay? In order to verify the solution obtained, evaluate all sums of money, including the required payment, at any date other than the present.

3-8. How much interest will $1,000 earn if it is invested at 6 per cent for 5 years?

3-9. A business firm contemplating the installation of labor-saving machinery has a choice between two different models. Machine A will cost $36,500, while machine B will cost $36,300. The repairs required for each machine are as follows:

Machine A: $1,500 at end of 5th year
           $2,000 at end of 10th year
Machine B: $3,800 at end of 9th year

The machines are alike in all other respects. If this firm is earning a 7 per cent return on its capital, which machine should be purchased?

Ans. Selecting the purchase date as our valuation date, a saving of $219 will accrue through purchase of machine B.

3-10. A fund was established in 1965, whose history is recorded below:

Deposit of $1,000 on 1/1/65
Deposit of $2,000 on 1/1/67
Withdrawal of $300 on 7/1/67
Deposit of $1,600 on 7/1/68
Withdrawal of $1,200 on 1/1/69

The fund earned interest at the rate of 3½ per cent compounded semiannually until the end of 1967. At that date, the interest rate was augmented to 4 per cent compounded semiannually. What was the principal in the fund at the end of 1971?

Ans. $3,855

3-11. If the sum of $2,000 is invested in a fund earning 7½ per cent interest compounded semiannually, what will be its final value at the end of 5 years?

Ans. $2,890

3-12. In the preceding problem, how long will it take the original sum of $2,000 to expand to $3,000?

3-13. On Apr. 1, 1970, an investor purchased stock of the XYZ Corp. at a total cost of $3,600. He then received the following semiannual dividends:

$105 on Oct. 1, 1970
$110 on Apr. 1, 1971
$110 on Oct. 1, 1971
$100 on Apr. 1, 1972

After receipt of the last dividend, the investor sold his stock, receiving $3,800 after deduction of brokerage fees. What semiannual rate did this individual realize on his investment?

Ans. 4.26 per cent per semiannual period, or 8.52 per cent compounded semiannually.

3-14. Calculate the effective rate corresponding to the following interest rates. Use logarithms to obtain the answer correct to the nearest tenth of one per cent.

5 per cent per annum compounded monthly
$6\frac{1}{2}$ per cent per annum compounded quarterly

3-15. An entrepreneur purchased a business for $40,000. As profits accumulated, he withdrew the following sums from the business:

$2,000 at the end of the first year
$2,600 at the end of the second year
$1,800 at the end of the third year

These withdrawals were invested in a mutual fund that earned 6 per cent per annum. At the end of the fourth year, the owner sold the business for $46,000. What was the value of the original capital at the termination of the venture? What was the average investment rate earned during the 4-year period?

3-16. A business firm plans to purchase a parcel of land and to build a structure on this land. However, since the structure is not required immediately, the firm is considering whether it should purchase the land and build the structure now or defer this action for 3 years. The current costs are as follows: land, $20,000; structure, $300,000.

The purchase price of the land and the cost of building the structure are expected to appreciate at the rate of 15 per cent and 4 per cent per annum, respectively. What will be the total cost of the land and structure 3 years hence?   *Ans.* $367,900

3-17. (This problem is presented for its mathematical interest.) Applying solely the principles of finance thus far developed, prove that $(1 + i)^4 = 1 + 4i + 6i^2 + 4i^3 + i^4$.

Solution: Consider that there are several funds, designated as funds A, B, C, etc., all earning the identical interest rate i. The sum of $1 is deposited in fund A. At the end of each period, the interest earned by fund A is transferred to fund B; the interest earned by fund B is transferred to fund C, etc. The left-hand member of this equation represents the value of the original sum of $1 at the end of the fourth period; the terms in the right-hand member represent, respectively, the principal in funds A, B, C, etc., at this date. This analysis can be applied as an alternative derivation of Eq. (3-2).

CHAPTER 4

SINKING FUNDS AND ANNUITIES—AMORTIZATION OF DEBT

4-1. Ordinary Sinking Fund. A sinking fund is an interest-earning fund in which equal deposits are made at equal intervals of time. Thus, if the sum of $100 is placed in a fund every 3 months, a sinking fund exists. A sinking fund is generally created for the purpose of gradually accumulating a specific sum of money required at some future date. For example, in Chap. 2 we stated that when a corporation has floated an issue of bonds it will often set aside a portion of its annual earnings to assure itself sufficient funds to retire the bonds at their maturity. The money thus held in reserve, however, need not remain idle but can be employed productively through investment in an interest-earning fund.

Assume that a business firm plans to purchase a new machine 5 years hence, the purchase price being $10,000. To accumulate this sum, it will make annual deposits in a sinking fund for the next 5 years, the fund earning interest at the rate of 4 per cent per annum. How much should each deposit be? (In reality, it would be pointless to deposit the fifth sum of money, since the fund will be closed at the same date. However, it will simplify our discussion if we consider that the deposit is actually made.)

Now, if the fund did not earn interest, the required annual deposit would be simply $10,000/5, or $2,000. However, since the principal in the fund will be continuously augmented through the accrual of interest, the actual deposit required will be somewhat less than $2,000. We shall soon derive a formula for calculating the required deposit.

In the discussion that follows, we shall assume, if nothing is stated to the contrary, that the following conditions obtain:

1. The sinking fund is established at the beginning of a specific interest period. We shall identify the date on which the fund is established as the "origin date" of the fund.

2. The interval between successive deposits, known as the "deposit period," is equal in length to an interest period.

3. Each deposit is made at the end of an interest period. Consequently, there is a time interval of one interest period between the date the fund is established and the date the first deposit is made.

4. The date at which a sinking fund is to terminate is always the last day of a specific interest period. We shall call this date the "terminal date" of the fund. The principal in the fund at its termination will, of course, include the interest earned during the last period and the final deposit in the fund, made on the terminal date.

A sinking fund satisfying the above requirements is referred to as an "ordinary sinking fund." The duration of the fund is referred to as the "term of the fund."

To illustrate the above definitions, assume that a sinking fund is created on Jan. 1, 1972, to consist of 10 deposits of $1,000 each. The interest (and deposit) period of the fund is 1 year. Then

Origin date = Jan. 1, 1972
Date of first deposit = Dec. 31, 1972
Terminal date (and date of last deposit) = Dec. 31, 1981
Term of fund = 10 years

At the end of each interest period, the principal in the fund increases abruptly as a result of the compounding of the interest earned during that period and the receipt of the periodic deposit. Where the principal in a sinking fund is to be calculated at a date intermediate between its origin and termination, we shall in all cases determine the principal on the last day of a particular interest period, immediately after these two events have transpired.

Example 4-1. The sum of $500 is deposited in a sinking fund at the end of each year for 4 years. If the interest rate is 6 per cent compounded annually, what is the principal in the fund at the end of the fourth year?

Solution. The development of the principal in the fund is recorded graphically in Fig. 4-1a by the method of chronological sequence. In constructing this graph, we have again assumed for convenience that interest is paid for fractional parts of an interest period. Moreover, in order to accentuate the general nature of this graph, we have constructed it on the basis of a 25 per cent interest rate rather than the actual rate of 6 per cent. This graph is intended solely as an aid in visualizing the growth of the principal, not as a means of achieving a graphical solution of the problem.

At the end of the first year, the sum of $500, represented by AB, is placed in the fund. During the second year, the principal earns interest at the rate of 6 per cent and amounts to $530 at the end of the second year (CD). A deposit

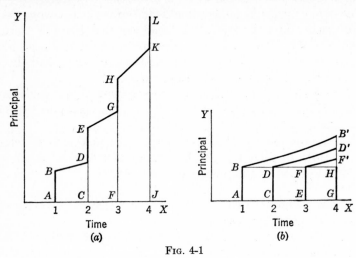

Fig. 4-1

of $500 at this time ($DE$) enhances the principal to the amount of $1,030 ($CE$). This two-cycle process is repeated each year until the final deposit KL is made. The growth of the principal is recorded in Table 4-1.

TABLE 4-1. SINKING-FUND PRINCIPAL

Year	Principal in fund at beginning	Interest earned	Deposit at end of period	Principal in fund at end
1	$ 0.00	$ 0.00	$500.00	$ 500.00
2	500.00	30.00	500.00	1,030.00
3	1,030.00	61.80	500.00	1,591.80
4	1,591.80	95.51	500.00	2,187.31

Thus, the principal in the sinking fund at the end of the fourth year is $2,187.31.

We shall now apply the method of separate evaluation to calculate the principal in the fund. In Fig. 4-1b, AB, CD, EF, and GH represent the deposits made at the end of their respective years. At the end of the fourth year, AB has expanded to GB', CD has expanded to GD', etc.   These values are computed in Table 4-2.

TABLE 4-2. VALUE OF DEPOSITS IN FUND

Deposit number	Amount	Number of years in fund	Value at the end of fourth year
1	$500.00	3	$500(1.06)^3 = \$\ 595.50$
2	500.00	2	$500(1.06)^2 = 561.86$
3	500.00	1	$500(1.06) = 530.00$
4	500.00	0	$500(1) = 500.00$
			Total $= \$2,187.30$

4-2. Equation for Sinking-fund Principal.   We shall now derive a general equation for the principal in a sinking fund.   Let

R = periodic deposit
i = interest rate
n = number of deposits made (= number of interest periods contained in the term of the fund)
S_n = principal in the sinking fund at the end of the nth period

To apply the method of chronological sequence, it is to be observed that the principal at the end of any period equals the principal at the end of the preceding period multiplied by the factor $(1 + i)$, and augmented by the periodic deposit R.   That is,

$$S_r = S_{r-1}(1 + i) + R$$

where r has any integral value less than n.   Hence, we obtain the following values:

$$S_1 = R$$
$$S_2 = R(1 + i) + R$$
$$S_3 = R(1 + i)^2 + R(1 + i) + R$$
$$\cdot\ \cdot\ \cdot\ \cdot\ \cdot\ \cdot\ \cdot\ \cdot\ \cdot\ \cdot\ \cdot\ \cdot\ \cdot\ \cdot\ \cdot\ \cdot\ \cdot\ \cdot$$
$$S_n = R(1 + i)^{n-1} + R(1 + i)^{n-2} + \cdots + R(1 + i) + R$$

For convenience, we shall reverse the sequence of the terms, thereby obtaining

$$S_n = R + R(1 + i) + R(1 + i)^2 + \cdots + R(1 + i)^{n-2} + R(1 + i)^{n-1}$$

The terms in the right-hand member of this equation constitute a geometrical series, in which the first term is R and the ratio of each term to the preceding term is $(1 + i)$. In Chap. 1, we derived the following equation for the sum of a geometrical series: If

$$S_n = a + ar + ar^2 + \cdots + ar^{n-2} + ar^{n-1}$$

then
$$S_n = a \frac{r^n - 1}{r - 1} \tag{1-9}$$

Applying this equation to the sinking-fund principal, but substituting R for a, and $(1 + i)$ for r, we obtain

$$S_n = R \frac{(1 + i)^n - 1}{(1 + i) - 1}$$

or
$$S_n = R \frac{(1 + i)^n - 1}{i} \tag{4-1}$$

This result can also be obtained by applying the method of separate evaluation, as shown in Table 4-3. By summing the values of all deposits, we obtain Eq. (4-1).

TABLE 4-3. VALUE OF DEPOSITS IN FUND

Deposit number	Amount	Number of periods in fund	Value of deposit at terminal date
1	R	$n - 1$	$R(1 + i)^{n-1}$
2	R	$n - 2$	$R(1 + i)^{n-2}$
.....	...	.....	....
$n - 1$	R	1	$R(1 + i)$
n	R	0	R

For the special case where the periodic deposit R is \$1, we shall use the symbol $s_{\overline{n}}$ to denote the principal in the fund at the end of the nth period. Hence,

$$s_{\overline{n}} = \frac{(1 + i)^n - 1}{i}$$

For the general case where the periodic deposit R has any value other than \$1, the principal S_n can be expressed in terms of $s_{\overline{n}}$ in this manner:

$$S_n = R s_{\overline{n}}$$

Values of $s_{\bar{n}}$ are listed in Appendix B for the usual interest rates encountered.

Example 4-2. A sinking fund consists of 15 annual deposits of $1,000 each, with interest earned at the rate of 4 per cent compounded annually. What is the principal in the fund at its terminal date?

Solution

$$R = \$1,000 \qquad i = 4 \text{ per cent} \qquad n = 15$$

From Table B-9, $s_{\overline{15}} = 20.02359$.

$$\therefore S_{15} = 1,000(20.02359) = \$20,023.59$$

In many problems, the principal in the sinking fund at its terminal date is the known quantity, and we must determine the periodic deposit required to accumulate this principal. Reversing Eq. (4-1), we obtain

$$R = S_n \frac{i}{(1 + i)^n - 1} \tag{4-2}$$

Let R'_n denote the periodic deposit required for the special case where S_n equals $1. Then

$$R'_n = \frac{i}{(1 + i)^n - 1} = \frac{1}{s_{\bar{n}}}$$

For the general case where the terminal principal S_n has any value other than $1, the periodic deposit R can be expressed in terms of R'_n in this manner:

$$R = S_n R'_n$$

Values of R'_n are listed in Appendix B.

Example 4-3. A corporation is establishing a sinking fund for the purpose of accumulating a sufficient capital to retire its outstanding bonds at maturity. The bonds are redeemable in 10 years, and their maturity value is $150,000. How much should be deposited each year if the fund pays interest at the rate of 3 per cent?

Solution

$$S_n = \$150,000 \qquad i = 3 \text{ per cent} \qquad n = 10$$

From Table B-7, $R'_{10} = 0.08723$.

$$\therefore R = 150,000(0.08723) = \$13,084.50$$

4-3. Alternative Derivation of Sinking-fund-principal Equation. With regard to the equation

$$S_n = R \frac{(1 + i)^n - 1}{i} \qquad (4\text{-}1)$$

it is interesting to observe that the numerator of the fraction in the right-hand member is the interest earned by \$1 for n periods at rate i, which we have designated as I_n. Hence, Eq. (4-1) can be written in the following alternative form:

$$S_n = R \frac{I_n}{i}$$

or

$$S_n = \frac{R}{i} I_n \qquad (4\text{-}3)$$

Equation (4-3) states, in effect, that the principal in the sinking fund at the end of the nth period equals the interest earned by the sum R/i for n periods. This equation, and consequently Eq. (4-1) as well, can also be derived by a method relying primarily on financial considerations alone. It is based on the following axiomatic principle: Assume there are two funds available for the investment of a given sum of money. If the two funds earn the same interest rate, then it is immaterial, as far as the total principal at any date is concerned, whether the entire sum is placed in one fund or divided in any proportion whatever between the two funds. Moreover, the total principal is not influenced by the transfer of any sum of money from one fund to the other at the end of an interest period.

Consider the sum R/i to be placed in fund A, which earns interest at the rate i per period. At the end of the first period, the interest earned is $(R/i)i$, or R. Instead of being retained in this fund, where it would be converted to principal, the interest earning R is withdrawn from fund A and deposited in fund B, which also earns interest at the rate i. Thus, at the commencement of the second period, the principal in fund A remains R/i. At the expiration of this period, the interest earned by fund A is again R; this is also withdrawn and deposited immediately in fund B. This cycle of events recurs for n periods. The result of siphoning off the interest earning from fund A to fund B is to maintain a constant principal in the former while steadily enhancing the principal in the latter. Since fund B receives periodic deposits of the constant amount R, it is actually a sinking fund; its principal immediately after the nth deposit has been made is designated by the symbol S_n.

If the interest earnings of fund A had been compounded rather than withdrawn from the fund, the principal P_n in fund A at the end of the nth period would have been

$$P_n = \text{original principal} + \text{interest earned}$$

or
$$P_n = \frac{R}{i} + \frac{R}{i} I_n$$

In accordance with the rule stated above for dual funds having identical interest rates, it is evident that the above sum represents the total principal in funds A and B.   Thus,

$$\text{Principal in fund A} + \text{principal in fund B} = \frac{R}{i} + \frac{R}{i} I_n$$

$$\frac{R}{i} + S_n = \frac{R}{i} + \frac{R}{i} I_n$$

$$\therefore S_n = \frac{R}{i} I_n \qquad (4\text{-}3)$$

Equation (4-3) can also be written in the following form:

$$\frac{S_n}{R} = \frac{I_n}{i} = \frac{I_n}{I_1} = \frac{\text{interest earned by \$1 in } n \text{ periods}}{\text{interest earned by \$1 in 1 period}}$$

4-4. The Sinking-fund-principal Curve.   Figure 4-2 depicts the growth of the principal S_n in a sinking fund by the method of chronological sequence.   In this diagram, AB, DE, GH, etc., represent the end-of-period deposits.   The ordinates at points B, E, H, etc., therefore represent the principal in the fund at the end of the first, second, third, etc., periods respectively.   Although the true graph is composed of a series of straight lines, we shall disregard this fact for convenience and connect the significant points O, B, E, H, etc., by a smooth curve, as shown.   This is tantamount to treating the equation

$$S_n = R \frac{(1 + i)^n - 1}{i} \qquad (4\text{-}1)$$

as if it applied to fractional as well as integral values of n.   We shall refer to the curve $OBEH$, etc., as the sinking-fund-principal curve.

We have found that, in accordance with Eq. (4-3), the sinking-fund principal S_n at the end of the nth period equals the interest earned in n periods by the sum R/i at interest rate i per period. Consequently,

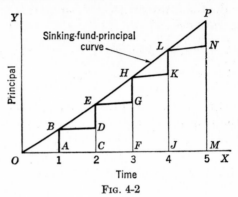

FIG. 4-2

curve $OBEH$, etc., can serve as both a sinking-fund-principal curve and an interest-time curve.

Consider three sums of money M_1, M_2, and M_3 to be deposited in the same fund at a particular date at an interest rate i. The appreciation of these sums with the passage of time is recorded by their respective principal-time graphs in Fig. 4-3. At the same date that these deposits are made, a sinking fund is established, also at interest rate i and consisting of periodic deposits of R each; this principal is represented by line (4). The relationship of the three sums of money to the periodic deposit R is such that

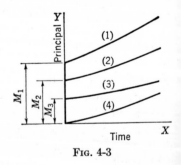

FIG. 4-3

$$M_1 > \frac{R}{i} \qquad M_2 = \frac{R}{i} \qquad M_3 < \frac{R}{i}$$

Now, applying Eq. (4-3), it is evident that M_1 acquires interest at a rate more rapid than the rate of increase of the sinking-fund principal. Hence, lines (1) and (4) diverge from one another. On the other hand, M_3 acquires interest at a slower rate, and lines (3) and (4) therefore converge toward one another until they intersect at some definite point. Finally, M_2 expands through the accrual of interest at the same rate at which the sinking-fund principal develops, which means that lines (2) and (4) are parallel to one another. The great significance of these graphical relationships will become apparent at a subsequent point, in connection with our study of perpetuities and amortization of debt.

4-5. Value of Sinking Fund at Origin Date. We have defined the quantity S_n as the principal in the sinking fund at the terminal date of the fund. It can also be defined simply as the value, at the terminal date, of the n sums of money deposited in the fund, based on an interest rate i. In many instances, however, it is also necessary to determine the value of these n sums of money at the origin date of the fund. This value will be denoted by the symbol $_nT_0$; the first subscript indicates the number of deposits to be made, and the second indicates that the sums are to be evaluated at the origin, or zero, date. We shall assume that the interest rate i of the fund equals the prevailing investment rate to be used in evaluating the n sums of money.

Now, both $_nT_0$ and S_n are the total value of the n deposits, but at different valuation dates. Therefore, these two quantities are equivalent to one another for a time interval of n interest periods. Applying Eq. (3-4), we obtain

$$_nT_0 = S_n(1 + i)^{-n}$$

But
$$S_n = R\,\frac{(1 + i)^n - 1}{i} \tag{4-1}$$

$$\therefore\ _nT_0 = R\,\frac{(1 + i)^n - 1}{i}\,(1 + i)^{-n}$$

$$= R\,\frac{1 - (1 + i)^{-n}}{i} \tag{4-4}$$

For the special case where the periodic deposit R is \$1, we shall use the symbol $_nt_0$ to denote the value of the sinking fund at its origin date. Hence,

$$_nt_0 = \frac{1 - (1 + i)^{-n}}{i}$$

For the general case where R has any value whatsoever, we can express $_nT_0$ in terms of $_nt_0$ in the following manner:

$$_nT_0 = R\,_nt_0$$

Values of $_nt_0$ are included in Appendix B.

Although we have defined $_nT_0$ with reference to a sinking fund, this quantity can be regarded as the value of the n given sums of money, based on an interest rate i, regardless of whether or not they are actually deposited in a sinking fund.

Example 4-4.   An individual is to receive the sum of $300 at the end of each year for 5 years.   One year prior to the receipt of the first sum, he decides to discount all five sums.   If the interest rate is 6 per cent, what proceeds will he obtain?

Solution

$$R = \$300 \qquad i = 6 \text{ per cent} \qquad n = 5$$

From Table B-13, $_5t_0 = 4.21236$.

$$\therefore \; _5T_0 = 300(4.21236) = \$1,263.71$$

In many problems, the value of the sinking fund at its origin date is the known quantity, and we must determine the periodic deposit R corresponding thereto.   Reversing Eq. (4-4), we obtain

$$R = \frac{_nT_0 i}{1 - (1 + i)^{-n}} \tag{4-5}$$

Let R_n'' denote the periodic deposit required for the special case where $_nT_0$ is $1.   Then

$$R_n'' = \frac{i}{1 - (1 + i)^{-n}} = \frac{1}{_nt_0}$$

For the general case where $_nT_0$ has any value whatsoever, R can be expressed in terms of R_n'' as follows:

$$R = {_nT_0}R_n''$$

Values of R_n'' are included in Appendix B.

There is a simple relationship between R_n' and R_n'', which can be obtained in the following manner:

$$R_n' = \frac{i}{(1 + i)^n - 1} \qquad \text{and} \qquad R_n'' = \frac{i}{1 - (1 + i)^{-n}}$$

Multiplying numerator and denominator of the fraction in the latter equation by $(1 + i)^n$, we obtain

$$R_n'' = \frac{i(1 + i)^n}{(1 + i)^n - 1}$$

$$\therefore R_n'' - R_n' = \frac{i(1 + i)^n - i}{(1 + i)^n - 1} = \frac{i[(1 + i)^n - 1]}{(1 + i)^n - 1} = i$$

or
$$R_n' + i = R_n'' \tag{4-6}$$

The equation for $_nT_0$ can also be written in alternate form, as follows:

$$_nT_0 = R\frac{1 - (1 + i)^{-n}}{i} = R\frac{D_n}{i} = \frac{R}{i}D_n \tag{4-7}$$

Hence, $_nT_0$ is the discount applicable to the sum R/i for n periods, based on an interest rate i. We also have

$$_nT_0 = R\frac{D_n}{i} = R\frac{D_n}{I_1}$$

$$\therefore \frac{_nT_0}{R} = \frac{D_n}{I_1} = \frac{\text{discount applicable to \$1 for } n \text{ periods}}{\text{interest earned by \$1 for 1 period}}$$

4-6. Value of Sinking Fund at Intermediate Date. In Fig. 4-4, line OC represents the principal in a sinking fund consisting of n deposits of R each at interest rate i. The ordinate BC is then the principal S_n at the terminal date. Using the same interest rate i, a principal-time curve is drawn through point C, intersecting the YY axis at point A. Hence,

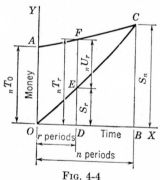

OA is the value of the sinking fund at the origin date. If we select some intermediate date, such as the end of the rth period, the ordinate DF of the principal-time curve represents the value of the sinking fund (i.e., equals the value of the n sums of money) at that date. We shall denote this quantity by $_nT_r$, the first subscript indicating the number of deposits constituting the fund and the second subscript indicating the valuation date. The ordinate DE of the sinking-fund-principal curve represents the principal in the fund at this date, or the value of the r sums already deposited in the fund. Evidently, then, the difference EF represents the value of the $(n - r)$ sums to be deposited in the future. This quantity will be denoted by $_nU_r$. (In the subsequent discussion, we shall replace the symbols $_nT_r$ and $_nU_r$ with the simpler forms T_r and U_r, the subscript n being understood.)

Fig. 4-4

The three variable quantities T_r, S_r, and U_r can be expressed both in terms of the periodic deposit R and in terms of the fixed quantities T_0 and S_n. To obtain the value of U_r, we can consider the actual sinking fund to be resolved into two distinct funds, the first comprising the r deposits of the past and the second comprising the $(n - r)$ deposits of the future. The end of the rth period is the terminal date of the first fund and the origin date of the second. We therefore have

$$S_r = R \frac{(1 + i)^r - 1}{i} \tag{4-8}$$

or

$$S_r = \frac{R}{i} I_r \tag{4-9}$$

$$U_r = R \frac{1 - (1 + i)^{-(n-r)}}{i} \tag{4-10}$$

or

$$U_r = R \,_{n-r}T_0 \tag{4-11}$$

or

$$U_r = \frac{R}{i} D_{n-r} \tag{4-12}$$

Then

$$T_r = S_r + U_r = R \frac{(1 + i)^r - (1 + i)^{-(n-r)}}{i} \tag{4-13}$$

or

$$T_r = \frac{R}{i} (I_r + D_{n-r}) \tag{4-14}$$

As proved above,

$$S_n = \frac{R}{i} I_n \tag{4-3}$$

and

$$T_0 = \frac{R}{i} D_n \tag{4-7}$$

Comparing the variable quantities to these fixed quantities, we have

$$\frac{S_r}{S_n} = \frac{I_r}{I_n} \qquad \text{or} \qquad S_r = S_n \frac{I_r}{I_n} \tag{4-15}$$

$$\frac{S_r}{T_0} = \frac{I_r}{D_n} \qquad \text{or} \qquad S_r = T_0 \frac{I_r}{D_n} \tag{4-16}$$

$$\frac{U_r}{S_n} = \frac{D_{n-r}}{I_n} \qquad \text{or} \qquad U_r = S_n \frac{D_{n-r}}{I_n} \tag{4-17}$$

$$\frac{U_r}{T_0} = \frac{D_{n-r}}{D_n} \qquad \text{or} \qquad U_r = T_0 \frac{D_{n-r}}{D_n} \tag{4-18}$$

$$T_r = T_0(1 + i)^r \tag{4-19}$$

or

$$T_r = S_n(1 + i)^{-(n-r)} \tag{4-20}$$

At the origin date, $T_0 = U_0$.

At the terminal date, $T_n = S_n$.

Example 4-5. A sinking fund consists of 20 end-of-year deposits of $1,000 each, the fund earning interest at the rate of $3\frac{1}{2}$ per cent. What is the value of the fund at the end of the seventh year?

Solution

$$R = \$1,000 \qquad i = 3\frac{1}{2} \text{ per cent} \qquad n = 20$$
$$r = 7 \qquad n - r = 13$$

By Table B-8, $s_{\overline{7}} = 7.77941$, and $_{13}t_0 = 10.30274$.

$$\therefore S_7 = 1,000(7.77941) = \$ 7,779.41$$
$$U_7 = 1,000(10.30274) = \underline{10,302.74}$$
$$T_7 = S_7 + U_7 = \underline{\$18,082.15}$$

Example 4-6. A corporation issued 200 bonds of $1,000 face value each, redeemable at par at the end of 15 years. To accumulate the funds required for redemption, the firm established a sinking fund consisting of annual deposits, the interest rate of the fund being 4 per cent. What was the principal in the fund at the end of the twelfth year?

Solution

$$S_n = 200 \times \$1,000 = \$200,000$$
$$i = 4 \text{ per cent} \qquad n = 15 \qquad r = 12$$

By Eq. (4-15),

$$S_{12} = S_{15} \frac{I_{12}}{I_{15}} = 200,000 \left(\frac{0.60103}{0.80094}\right)$$
$$= \$150,081$$

4-7. Special Problems Pertaining to Sinking Funds. In the solution of numerical problems in finance, it often occurs that the most obvious method of solution is not the most expedient one. A brief analysis of each problem will generally disclose an alternative method of solution that greatly curtails the arithmetical calculations required. We shall illustrate this by solving two special problems regarding sinking funds.

Example 4-7. A business firm established a sinking fund for the purpose of replacing an existing asset at the end of 10 years. The fund consisted of semi-annual payments of $600 each and earned interest at the rate of 4 per cent compounded semiannually. At the terminal date, however, the asset was still in good operating condition, and the principal accumulated in the fund was allowed to remain there. What was the principal in the fund at the end of 13 years (i.e., 3 years after the terminal date of the sinking fund)?

Solution. The growth of the principal in the fund is represented graphically in Fig. 4-5a. At the terminal date, the principal in the fund has the value AB. The sinking fund is discontinued at this date, and the principal developed then

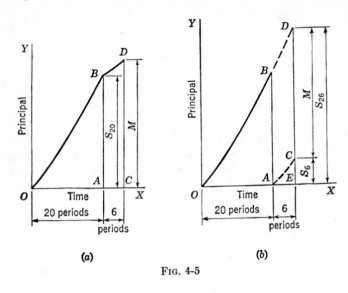

(a)   (b)

Fig. 4-5

continues to expand by the simple money-time relationship, Eq. (3-2). Let M denote the principal at the end of the thirteenth year. We then have

$$\text{No. of deposits in sinking fund} = 20 \quad i = 2 \text{ per cent}$$
$$AB = S_{20} = 600s_{\overline{20}} = 600(24.29737) = \$14,578.42$$
$$M = CD = AB(1.02)^6 = 14,578.42(1.12616)$$
$$= \$16,417.62$$

An alternative method of solution is the following: Assume that deposits were actually made in the sinking fund for the entire 13-year period under consideration. For this condition, the principal in the sinking fund at the end of the thirteenth year is represented by line ED in Fig. 4-5b. However, since in reality the deposits were discontinued after the tenth year, we must deduct from ED the value at that date of the six hypothetical deposits. This value is represented by EC. We then have

$$ED = 600s_{\overline{26}|}$$
$$EC = 600s_{\overline{6}|}$$

$$M = ED - EC = 600(s_{\overline{26}|} - s_{\overline{6}|}) = 600(33.67091 - 6.30812) = 600(27.36279)$$
$$= \$16,417.67$$

This method of solution is considerably simpler than the first, since it circumvents the need for multiplying odd figures.

Example 4-8. A corporation planned to construct an additional plant at the end of 10 years for an estimated cost of \$100,000. To accumulate this sum, it

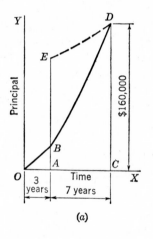 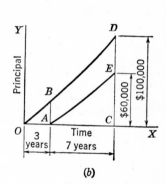

(a) (b)

Fig. 4-6

established a sinking fund consisting of annual deposits, the fund earning interest at the rate of 4 per cent. At the end of the third year, however, the firm decided to erect a larger plant than originally contemplated, with the cost now estimated at \$160,000. What should be the amount of the annual deposit for the remaining 7 years?

Solution. Figure 4-6a records the growth of the principal in the fund. There are two distinct sinking-fund-principal curves, curve OB for the first 3 years and curve BD for the remaining 7 years.

Several methods are available for determining the periodic deposit R applicable to the last 7 years, one of which is the following: Selecting the end of the third year as our valuation date, we can evaluate both the principal AB in the sinking fund and the sum AE, which at that date is equivalent to \$160,000 at the end of the tenth year. The difference BE between these two sums represents the value of the remaining seven deposits.   We then have, by Eq. (4-15),

$$AB = S_3 = S_{10}\frac{I_3}{I_{10}} = 100{,}000\left(\frac{0.12486}{0.48024}\right)$$
$$= \$26{,}000$$
$$AE = 160{,}000(1.04)^{-7} = 160{,}000(0.75992)$$
$$= \$121{,}587.20$$
$$BE = AE - AB = \$95{,}587.20$$
$$R = 95{,}587.20R_7'' = 95{,}587.20(0.16661)$$
$$= \$15{,}925.67$$

An alternative, and much simpler, method of solution is the following: In Fig. 4-6b, assume that the original sinking fund remained unaltered, with the identical deposit being made periodically for the full 10 years.   If this were true, the principal at the end of the tenth year would be $100,000 (*CD*).   It would thus be necessary to establish at the end of the third year a second sinking fund, capable of developing a principal of $60,000 (*CE*) at the expiration of the next 7-year period.   The true annual deposit required during the last 7 years will then be the sum of the payments applying to these two separate funds.   Let *A* denote the deposit pertaining to the original fund and *B* the deposit required by the second fund.   We then have

$$A = 100{,}000R_{10}' = \$\ 8{,}329.00$$
$$B = 60{,}000R_7' \quad = \quad 7{,}596.60$$
$$R = A + B \quad = \$15{,}925.60$$

On the basis of these two numerical problems, it is possible to draw the following conclusion: In a problem of finance in which there occurs a discontinuity of a given cycle of events, the simplest method of solution is achieved by assuming that the given cycle of events actually persisted during the entire interval of time involved in the problem and then applying whatever correction is required to reflect the true state of affairs.   We shall refer to this as the "rule of continuity."

In the sinking-fund problems treated thus far, the fund has been established for the purpose of accumulating one specific sum of money by a particular date in the future.   However, there is no necessity to restrict the problem in this manner; a fund can be established to accumulate several sums of money by various future dates.

Example 4-9.   A business firm must make a disbursement of $5,000 on Dec. 31, 1984, and of $8,000 on Dec. 31, 1985.   To accumulate these sums, a sinking fund will be established on Jan. 1, 1980, into which deposits will be made annually.

The interest rate of the fund is 4 per cent.   What is the required amount of the annual deposit?

Solution

$$n = 6 \qquad i = 4 \text{ per cent}$$

At any valuation date we select, the value of the six deposits must equal the combined value of the two disbursements.   The most convenient date to select is the end of 1985, which is the terminal date of the fund.   We then have

$$S_6 = 5,000(1.04) + 8,000$$
$$= \$13,200$$
$$R = 13,200R_6' = 13,200(0.15076)$$
$$= \$1,990.03$$

That the periodic deposit of this sum will provide the necessary funds can be verified in the following manner:

Principal in fund at end of 1984 $= S_5 = 1,990.03(5.41632)$	$=$	\$10,778.64
Disbursement	$=$	5,000.00
Balance on Jan. 1, 1985	$=$	\$ 5,778.64
Interest earned in 1985	$=$	231.15
Final deposit	$=$	1,990.03
Principal in fund on 12/31/85	$=$	\$ 7,999.82

The slight inaccuracy of the result arises from the limits of the accuracy of the tables.

4-8. Extension of Definitions.   In our study of sinking funds, we have dealt with n sums of money to be deposited in a fund under the following conditions:

1. Each sum, at its date of deposit, has the value R.
2. The interval between successive deposits is one interest period.
3. Each deposit is made at the end of an interest period.

With the assumption that the rate of interest earned by this fund is identical with the investment rate prevailing in the financial market, we have derived equations to enable us to evaluate these n sums of money at any given date.   However, by virtue of our basic assumptions regarding the investment of money, we have postulated that the mathematical relationship governing the variation of money with time applies to every sum of money, regardless of whether that sum is deposited in a fund, expended, or hoarded.   As a result, the terms used with reference to sinking funds can be extended to apply simply to n sums of money, the sums being of such amount that at the end of each interest period one particular sum has the value R, based on an interest rate i.

As an illustration, consider the quantity $_nT_0$.   With reference to a sinking fund, this was defined as the value at the origin date of the fund of the n sums of money to be placed in the fund.   However, by deleting the qualification that these sums are to be deposited, we can now generalize our definition of $_nT_0$ in the following manner: Consider n sums of money, of differing amounts.   These sums are to be evaluated at the beginning of a particular interest period, based on an interest rate i.   The first sum will have the value R at the end of the first period; the second sum will have the value R at the end of the second period; the nth sum will have the value R at the end of the nth period.   The total value of these n sums at the stipulated valuation date is $_nT_0$.   The final disposition of these sums of money is of no consequence in calculating the value $_nT_0$, except insofar as this affects the interest rate to be used in the calculations.

4-9. Annuities.   A series of equal payments made at equal intervals of time constitutes an "annuity."   Thus, the periodic payment of rent, the periodic payment of interest to bondholders, the annual payment of a fixed sum of money for the use of a patent, etc., are all illustrations of annuities.   (In the case of monthly disbursements, these are also considered to be annuities, even though a month is not, strictly speaking, a fixed period of time.)

In the material that follows, we shall assume that the following conditions apply to an annuity:

1. The interval between successive payments is equal in length to an interest period.

2. Each payment is made at the end of an interest period.

An annuity satisfying these requirements is referred to as an "ordinary annuity."   We shall designate the beginning of the first interest period of the term of the annuity as its "origin date."   Hence, one interest period intervenes between the origin date of an annuity and the date of the first payment.   The end of the last interest period, or the date on which the last payment is made, will be called the "terminal date" of the annuity.   If the annuity consists of n payments, there are n interest periods constituting the term of the annuity.   Where the value of an annuity is to be determined at an intermediate date, we shall in all cases calculate this value at the end of an interest period, immediately after the payment for that period has been made.

It is evident at once that the terms used with reference to a sinking fund, after having been extended in scope as was done with $_nT_0$, can likewise be applied to the case of annuities. It is merely necessary to know the prevailing investment rate at which the payments are to be evaluated. We thus have the following definitions:

1. S_r is the value at the end of the rth period of the r payments already made, including the payment made at the end of the rth period.

2. $_nU_r$ is the value at the end of the rth period of the $(n - r)$ payments to be made in the future.

3. $_nT_r$ is the value at the end of the rth period of all n payments constituting the annuity, or

$$_nT_r = S_r + {}_nU_r$$

We have designated the graph representing the principal in a sinking fund as a sinking-fund-principal curve. When applied to an annuity, this graph represents the value of the payments made prior to and on a given date, a value which is the quantity S_r. We shall therefore refer to this graph as an annuity-principal curve when it pertains to an annuity.

Example 4-10. On Jan. 1, 1971, there was a balance in a bank account of $6,000. Withdrawals of $500 each were made at the end of each year for 4 years. If the account earned interest at the rate of 4 per cent compounded annually, what was the balance in the account at the end of 1974, immediately after the last withdrawal was made?

Solution. The history of the principal in the account is recorded in Fig. 4-7a. OA represents the principal of $6,000 at the beginning of 1971. At the end of that year, through the accrual of interest, this amount has expanded to BC. A withdrawal of $500 on that date (CD) depresses the principal to BD. This cycle

TABLE 4-4. PRINCIPAL IN BANK ACCOUNT

Year	Principal at beginning	Interest earned	End-of-year withdrawal	Principal at end
1971	$6,000.00	$240.00	$500.00	$5,740.00
1972	5,740.00	229.60	500.00	5,469.60
1973	5,469.60	218.78	500.00	5,188.38
1974	5,188.38	207.54	500.00	4,895.92

is repeated each year until at the end of 1974 the balance LN remains. This value is calculated in Table 4-4, based on the method of chronological sequence. As shown in the table, the principal at the end of 1974 is \$4,895.92.

Applying the method of separate evaluation, we can determine the principal in the account at the required date by evaluating separately both the original principal in the account and the four periodic withdrawals. Since the latter constitute an annuity, their value at the end of 1974 can be calculated by applying the equation for S_n. The interest rate to be used in this calculation is, of course, 4 per cent, the rate which the money earned while in the account.

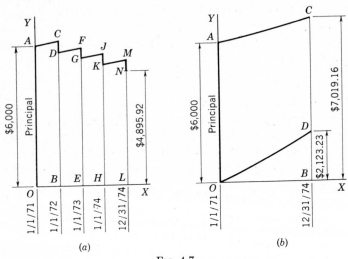

FIG. 4-7

Figure 4-7b is a graphical representation of this problem based on the method of separate evaluation. OA again represents the original principal of \$6,000. This amount, if left intact, would have expanded along the path AC, amounting to BC at the expiration of 4 years. Curve OD is the annuity-principal curve, its ordinate representing the value of the withdrawals made up to a given date. At the end of 1974, which is the terminal date of the annuity, the value of the four withdrawals is represented by BD. The line DC therefore represents the residue in the account. We then have

$$BC = 6,000(1.04)^4 = 6,000(1.16986) = \$7,019.16$$
$$BD = 500s_{\overline{4}|} \qquad = 500(4.24646) \quad = \underline{2,123.23}$$
$$DC = \text{balance in account} \qquad\qquad = \underline{\$4,895.93}$$

Example 4-11. A business firm must make a disbursement of $800 at the end of each year from 1974 to 1977 inclusive. In order to meet the obligations as they fall due, it deposits in a fund at the beginning of 1974 an amount of money that will be just sufficient to provide the necessary payments. If the interest rate is 3 per cent compounded annually, what sum should be deposited?

Solution. The history of the principal in the account is recorded in Fig. 4-8*a*. The deposit OA must be of such magnitude that the last withdrawal LM will close the account; or, to express this geometrically, point M will fall on the XX axis. This problem is also represented graphically in Fig. 4-8*c*, based on the

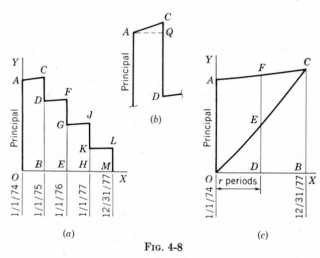

Fig. 4-8

method of separate evaluation. OA again represents the principal to be placed in the fund at the beginning of 1974, the magnitude of which we must determine. If this principal were left unimpaired, it would expand, through the accrual of interest, along the path AC and amount to BC at the end of 1977. The four withdrawals made from the fund constitute an annuity, whose origin date is the beginning of 1974 and whose terminal date is the end of 1977. This annuity is to be evaluated at the interest rate of 3 per cent. In Fig. 4-8*c*, the value of the withdrawals made up to a given date is represented by the annuity-principal curve OC. The difference between the ordinates of the two curves at a given date represents the balance in the account at that date. At the terminal date of the annuity, the curves AC and OC must intersect.

Let X denote the sum to be deposited in the fund on Jan. 1, 1974. The controlling relationship is

$$\text{Value of deposit} = \text{value of annuity}$$

This relationship obtains for any valuation date we select; the most convenient date to use is obviously the beginning of 1974, which is both the date of the deposit and the origin date of the annuity.   Then

$$X = R\,_nt_0$$
$$R = \$800 \qquad i = 3 \text{ per cent} \qquad n = 4$$
$$X = 800(3.71710)$$
$$= \$2,973.68$$

Curve AC in Fig. 4-8c represents not only the value of the original deposit but the value of the annuity as well, since the two are equal.   At some intermediate date, such as the end of the rth period, the value of the original deposit and of the annuity is $DF(=\,_nT_r)$, while the value of the withdrawals made up to that date is $DE(=S_r)$. EF, the difference between the two, represents both the value of the future payments and the principal remaining in the account $(=\,_nU_r)$.   From the geometry of this diagram, it is evident that as time elapses the principal U_r decreases at an ever-increasing rate; that is, the disintegration of the principal is accelerated.

The rate of decrease of the principal in the account can also be analyzed by applying the method of chronological sequence.   Fig. 4-8a depicts the exact history of this principal; a portion of this diagram has been reproduced in Fig. 4-8b but magnified for convenience. QC represents the interest earned during the first year, which is 3 per cent of \$2,973.68, or \$89.21. CD represents the \$800 withdrawal made at the end of the year.   This withdrawal has the effect of absorbing the interest earning of \$89.21 (QC) and depressing the original principal OA by \$710.79 (QD).   To express this in another manner, we can consider a portion of each end-of-year withdrawal to be supplied by the interest earned during the year, the balance to be supplied by a reduction of principal.   Since the principal, and therefore the interest earning, diminishes with each successive period, it follows that the portion of the withdrawal supplied by a reduction of principal becomes progressively larger.   Consequently, as time elapses, the principal decreases at a constantly increasing rate.   This fact is emphasized if we connect the significant points A, D, G, K, and M of Fig. 4-8a with a smooth curve.

Example 4-12.   (P.E. examination problem.)   A father wishes to develop a fund for his newborn son's college education.   The fund is to pay \$2,000 on the 18th, 19th, 20th, and 21st birthdays of the son.   The fund will be built up by the deposit of a fixed sum on the son's 1st to 17th birthdays, inclusive.   If the fund earns 2 per cent, what should the yearly deposit into the fund be?

Solution.   In Fig. 4-9, curve OB represents the principal in the sinking fund, and curve CB represents the total value of the fund.   The sinking fund is discontinued at the end of the 17th year. If the principal in the fund were left intact, it would expand along the principal-time curve BE, which is a continuation of curve CB. The value of the withdrawals made from the fund up to a given date is represented by the annuity-principal curve AE. Curves BE and AE intersect at the end of the 21st year. The controlling relationship is

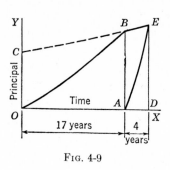

FIG. 4-9

Value of sinking fund = value of annuity

Let R denote the periodic deposit in the fund. Selecting the end of the 17th year as our valuation date, we have

$$Rs_{\overline{17}} = 2{,}000 \; {}_4t_0$$
or
$$R = 2{,}000 \; {}_4t_0 \, R'_{17}$$
$$= 2{,}000(3.80773)(0.04997)$$
$$= \$380.55$$

As an alternative solution, we can select the origin date of the sinking fund as our valuation date. To calculate the value of the annuity at this date, we can apply the rule of continuity by assuming that withdrawals were actually made for the entire 21-year period and then applying the necessary correction. This gives

$$\text{Value of annuity at son's date of birth} = 2{,}000 \; ({}_{21}t_0 - {}_{17}t_0)$$
$$= 2{,}000(17.01121 - 14.29187)$$
$$= \$5{,}438.68$$
$$\therefore R = 5{,}438.68 R''_{17} = 5{,}438.68(0.06997)$$
$$= \$380.54$$

Example 4-13. The sum of \$1,000 was deposited in a fund at the end of each year for 6 consecutive years, the first deposit having been made at the end of 1968. Withdrawals of uniform amount are to be made at the end of each year for 4 consecutive years, the first withdrawal to be made at the end of 1983. The balance in the fund is to be \$600 immediately after the last withdrawal is made. If the interest rate of the fund is 5 per cent per annum, what is the amount of the periodic withdrawal, to the nearest dollar?

Solution. Let W denote the periodic withdrawal. Figure 4-10 shows the dates of the deposits and withdrawals.

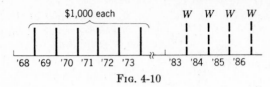

$$\text{F{\scriptsize IG}}. \ 4\text{-}10$$

The most suitable valuation date is the end of 1986. By assuming that the deposits continued up to this date and then applying the necessary correction, we obtain

$$\text{Value of deposits} = 1{,}000(s_{\overline{19}} - s_{\overline{13}})$$
$$= 1{,}000(30.53900 - 17.71298)$$
$$= \$12{,}826$$
$$\text{Value of withdrawals} = \$12{,}826 - \$600 = \$12{,}226$$
$$W = 12{,}226R'_4 = 12{,}226(0.23201) = \$2{,}837$$

4-10. Amortization of Loans. In Art. 3-10, we found that it is possible to devise virtually any system of payments for the purpose of liquidating a given debt. One arrangement that is very frequently employed in the case of many long-term loans, such as mortgage loans, is to have the debt discharged by a uniform series of payments (i.e., an annuity) extending over the duration of the loan. A loan that is repaid in this manner is said to be "amortized."

If nothing is stated to the contrary, we shall assume that the payment period (the interval between successive payments) is equal in length to one interest period and that the interval between the date the loan is consummated and the date the first payment is made is one payment, or interest, period. On this basis, each payment is made at the end of an interest period, and the date of the loan is also the origin date of the annuity.

In order to determine the periodic payment required to liquidate the debt, it is simply necessary to equate the amount of the loan with the value of the annuity at its origin date.

Example 4-14. A loan of \$5,000 is to be amortized by means of four equal payments at the end of each year, the first payment to be made 1 year after the loan is consummated. If the loan earns interest at the rate of 6 per cent per annum, what is the amount of the annual payment?

Solution

$$_nT_0 = \$5,000 \qquad i = 6 \text{ per cent} \qquad n = 4$$
$$R = 5,000R_4'' = 5,000(0.28859)$$
$$= \$1,442.95$$

The accuracy of this result can be verified by tracing the history of the loan, as shown in Table 4-5.

TABLE 4-5. PRINCIPAL OF LOAN

Year	Principal of loan at beginning	Interest for year	End-of-year payment	Principal of loan at end
1	$5,000.00	$300.00	$1,442.95	$3,857.05
2	3,857.05	231.42	1,442.95	2,645.52
3	2,645.52	158.73	1,442.95	1,361.30
4	1,361.30	81.65	1,442.95	0.00

Figure 4-11*a* represents the variation of the principal of the loan.   The value of the loan at its commencement, $5,000, is represented by OA.   At the end of the

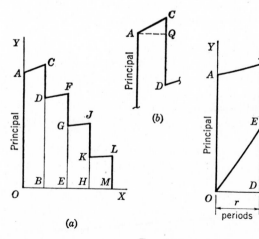

FIG. 4-11

first year, through the earning of interest, the principal has expanded to BC. The first payment of \$1,442.95, made at the end of the first year and represented by CD, reduces the principal to BD. This two-cycle process—enlargement of the principal of the loan through the earning of interest and reduction of the principal by the end-of-year payment—is repeated each year until at the end of the fourth year the debt has been fully liquidated.

In Fig. 4-11b, the history of the fund during the first year is represented with a larger scale. QC represents the interest of \$300 earned by the loan during the first year. The end-of-year payment CD can be regarded as consisting of two elements, as follows:

1. A portion CQ to pay the interest that accrued on the loan during that year.
2. A portion QD to reduce the principal of the loan.

Since the principal, and therefore the annual interest earning, diminishes with each successive period, it follows that the portion of the periodic payment applicable toward the reduction of principal becomes progressively larger. Consequently, as time elapses, the balance of the loan decreases at an ever-increasing rate. To express this in another manner, the liquidation of the debt is accelerated as time elapses.

Figure 4-11c also represents the amortization of the loan, based on the method of separate evaluation. OA again represents the \$5,000 loan at its commencement, and curve AC represents the value of this sum at all subsequent dates. Each ordinate of this curve indicates the value to which the principal of the loan would have grown by that particular date if no payments had been made. The annuity-principal curve OC represents the value of the payments made up to a given date, based on an interest rate of 6 per cent. The two curves intersect when n equals 4 years. If we select some intermediate date, such as the end of the second year, then $DF(=\,_4T_2)$ is the value of the original loan; $DE(=S_2)$ is the value of the two payments already made. The difference between these ordinates $EF(=\,_4U_2)$ is the balance of the loan at that date, or the value of the future payments. It is evident in this diagram also that the principal of the loan decreases at a steadily increasing rate with the passage of time.

Example 4-15. (P.E. examination problem.) A \$10,000 mortgage is being amortized by means of 20 equal yearly installments at an interest rate of 6 per cent. The agreement provides for paying off the mortgage in a lump sum at any time with an amount equal to the unpaid balance plus a charge of 1 per cent of the unpaid balance. What would have to be paid to discharge the mortgage after 10 payments have been made?

Solution. The unpaid balance of the mortgage at an intermediate date is the quantity $_nU_r$.

$$_nT_0 = \$10,000 \qquad n = 20 \qquad r = 10 \qquad n - r = 10 \qquad i = 6 \text{ per cent}$$

By Eq. (4-18),

$$U_r = T_0 \frac{D_{n-r}}{D_n}$$

$$U_{10} = 10,000 \left[\frac{1 - (1.06)^{-10}}{1 - (1.06)^{-20}} \right] = 10,000 \left(\frac{0.44161}{0.68820} \right)$$

$$= \$6,416.88$$

1 per cent of $U_{10} = \underline{\qquad 64.17}$

Payment required $= \underline{\$6,481.05}$

4-11. Determining the Interest Rate of an Annuity.

In many problems pertaining to annuities that arise in practice, the known quantities are the value of the annuity at either the origin or terminal date and the amount of the periodic payment, while the interest rate by which the given quantities are related is unknown. This is often true, for example, where an asset is purchased on the installment plan. The purchaser knows the purchase price of the asset and the periodic payment he is obligated to make, but he is not directly aware of the interest rate implicit in the loan. Only an approximate solution of this type of problem is possible, and a solution by straight-line interpolation yields results of a sufficient degree of accuracy.

Example 4-16. An asset costing \$5,000 was purchased on the installment plan, the terms of sale requiring that the buyer make a down payment of \$2,000 and five annual payments of \$720 each. The first of these periodic payments is to be made 1 year subsequent to the date of purchase. What is the interest rate pertaining to this loan?

Solution

$$_nT_0 = \$3,000 \qquad n = 5 \qquad R = \$720$$

$$_5t_0 = \frac{3,000}{720} = 4.16667$$

By Appendix B, we find

$$_5t_0 \text{ for } i = 6 \text{ per cent is } 4.21236$$
$$_5t_0 \text{ for } i = 6\tfrac{1}{2} \text{ per cent is } 4.15568$$

By straight-line interpolation between these two values,

$$i = 6.4 \text{ per cent}$$

4-12. Sinking-fund Period Different from Interest Period. In our study of sinking funds, we have thus far confined ourselves to ordinary funds satisfying the following two requirements:

1. The deposit period is identical with the interest period.
2. The deposits are made at the end of each deposit period.

There are, however, many funds that depart from the payment sequence listed above.   We shall first consider those funds in which only the second condition prevails; i.e., deposits are made at the end of each deposit period, but this period does not coincide with the interest period.   In all cases, we shall assume that the deposit period is an exact multiple of the interest period and that both these periods commence on the same day of the year.   Let

R = amount of the periodic deposit
i = interest rate
n = number of deposits made
m = number of interest periods contained in one deposit period
S'_n = principal in the fund at the end of the nth deposit period

This special sinking fund can be transformed to an ordinary fund by the simple expedient of expanding the interest period to coincide in length with the deposit period.   This will require replacing the interest rate i with an equivalent rate applicable to this enlarged period.   Let r denote this equivalent rate.   Then

$$r = (1 + i)^m - 1$$

and

$$1 + r = (1 + i)^m$$

We can now apply Eq. (4-1), but with r replacing i:

$$S'_n = R \frac{(1 + r)^n - 1}{r}$$

Substituting the value of r, we obtain

$$S'_n = R \frac{(1 + i)^{mn} - 1}{(1 + i)^m - 1} \tag{4-21}$$

This equation can also be written in the following form:

$$\frac{S'_n}{R} = \frac{(1 + i)^{mn} - 1}{(1 + i)^m - 1} = \frac{I_{mn}}{I_m}$$

$$= \frac{\text{interest earned by \$1 in } n \text{ deposit periods}}{\text{interest earned by \$1 in 1 deposit period}}$$

In Eq. (4-21), consider the numerator and denominator of the fraction to be divided by i. This gives

$$S'_n = \frac{R s_{\overline{mn}}}{s_{\overline{m}}} \tag{4-22}$$

Similarly, if $_n T'_0$ denotes the value of the sinking fund at its origin date, we obtain

$$_n T'_0 = R \frac{1 - (1 + i)^{-mn}}{(1 + i)^m - 1} \tag{4-23}$$

Dividing numerator and denominator of this fraction by i produces

$$_n T'_0 = \cdot \frac{R \,_{mn} t_0}{s_{\overline{m}}} \tag{4-24}$$

Example 4-17. Deposits of $1,500 each were made in a sinking fund at the end of each year. If the fund earned interest at the rate of 6 per cent per annum compounded quarterly, what was the principal in the fund immediately after the seventh deposit was made?

Solution

$$R = \$1,500 \qquad n = 7 \qquad m = 4 \qquad mn = 28 \qquad i = 1\tfrac{1}{2} \text{ per cent}$$

$$S'_n = R \frac{s_{\overline{mn}}}{s_{\overline{m}}} = 1,500 \frac{s_{\overline{28}}}{s_{\overline{4}}} = 1,500 \left(\frac{34.48148}{4.09090} \right)$$

$$S'_7 = \$12,643.23$$

4-13. Sinking Funds with Beginning-of-period Deposits. In the case of an ordinary sinking fund, the fund is established at the beginning of an interest period, while the periodic payment is made at the end of an interest period. Consequently, the first deposit is made one interest period after the origin date, and the final deposit is made on the terminal date. This sequence of deposits is represented in Fig. 4-12a.

Consider now the case of a sinking fund which is established at the beginning of an interest period, with the first deposit made immediately. The second deposit is then made at the commencement of the second interest period, etc. If n deposits are made in the fund and n interest periods constitute the term of the fund, then the final deposit is made one interest period prior to the terminal date. This sequence of deposits is represented in Fig. 4-12b.

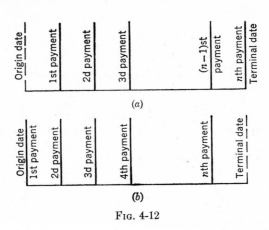

Fig. 4-12

Let $_nT_0''$ denote the value of this fund at its origin date and S_n'' denote the value at its terminal date. Assume first that the initial deposit, made on the origin date, had been omitted. The value of the fund at this date would then be the total value of the remaining $(n - 1)$ deposits, as given by Eq. (4-4). To obtain the true value of the fund, we then add the initial deposit. This procedure yields

$$_nT_0'' = R\left[\frac{1 - (1 + i)^{-(n-1)}}{i} + 1\right] \tag{4-25}$$

or
$$_nT_0'' = R(_{n-1}t_0 + 1) \tag{4-25a}$$

To obtain the value of S_n'', we next assume that a deposit was actually made at the terminal date. The principal in the fund would then be the total value at this date of the $(n + 1)$ deposits, as given by Eq. (4-1). To obtain the true value of S_n'', we then deduct this hypothetical deposit.

This procedure yields

$$S_n'' = R\left[\frac{(1 + i)^{n+1} - 1}{i} - 1\right] \tag{4-26}$$

or

$$S_n'' = R(s_{\overline{n+1}} - 1) \tag{4-26a}$$

4-14. Perpetuities. Consider that a sum of money M is deposited in a fund at the beginning of a particular interest period, the fund earning interest at the rate i per period. Withdrawals of R each are made from the fund at the end of each interest period. These withdrawals therefore constitute an annuity. At the end of the nth period, the principal in the fund equals the value of the original deposit, less the value of the n withdrawals made up to that date. But

$$\text{Value of original deposit} = M + MI_n$$

$$\text{Value of past withdrawals} = \frac{R}{i} I_n$$

$$\therefore \text{Principal in fund} = M + MI_n - \frac{R}{i} I_n$$

$$= M + \left(M - \frac{R}{i}\right) I_n$$

The relationship of M to R/i therefore determines whether the principal will increase, decrease, or remain constant. The principal in the fund is represented in Fig. 4-13 for three distinct cases by the method of separate evaluation. In a, the deposit M exceeds R/i. Consequently, the original deposit appreciates in value more rapidly than does the value of the withdrawals made, and the two curves therefore diverge from one another, producing an increase of principal. On the other hand, in case b, the deposit M is less than R/i; the two curves therefore converge toward one another until a point is reached at which the principal vanishes. Finally, in case c, M equals R/i; the two curves are therefore parallel, and the principal remains constant.

These three cases are also represented in Fig. 4-14 by the method of chronological sequence. Where M exceeds R/i, the interest earnings exceed the withdrawals, and the principal therefore expands. Where M is less than R/i, the interest earnings are less than the withdrawals, and the principal is impaired until it is completely dissipated. Finally, where

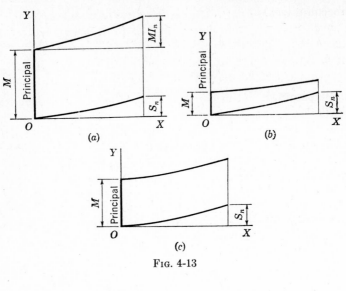

Fig. 4-13

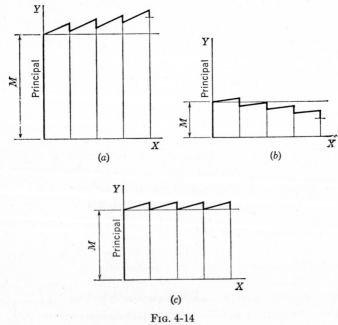

Fig. 4-14

M equals R/i, each interest earning exactly balances the periodic withdrawal, and the original principal is left unaltered. Thus, in the last case, the sum M deposited in the fund is capable of providing an indefinite number of periodic withdrawals of R each.

There are many annuities that are considered to continue indefinitely; that is, the number of payments is assumed to be infinite. Such an annuity is termed a "perpetuity." To provide for a perpetuity, it is necessary to deposit in a fund a sum of money of such magnitude that the interest earned between successive payments exactly equals the amount of the periodic payment, thereby maintaining the original principal permanently intact.

An endowment furnishes a simple illustration of a perpetuity. Assume that an individual wishes to establish a scholarship for a university of $5,000 annually and that the fund in which the money is to be held earns 4 per cent interest per annum. To earn $5,000 interest annually, the sum to be deposited is

$$M = \frac{R}{i} = \frac{5,000}{0.04} = \$125,000$$

This sum must be deposited 1 year prior to the date of the first payment.

In the above example, the payment period coincided with the interest period, but this is not necessarily the case. In general, let

R = periodic payment of a perpetuity
i = interest rate
m = number of interest periods contained in one payment period
M = principal to be deposited in the fund one payment period prior to the date of the first payment

Equating the periodic payment R to the interest earned by M in one payment period ($= m$ interest periods), we obtain

$$M[(1 + i)^m - 1] = R$$

or

$$M = \frac{R}{(1 + i)^m - 1} \tag{4-27}$$

To obviate the necessity of dividing by an odd figure, the formula for M can be transformed by dividing numerator and denominator of the fraction by i, thereby obtaining

$$M = \frac{R/i}{[(1+i)^m - 1]/i} = \frac{R}{i}\frac{i}{(1+i)^m - 1}$$

or

$$M = \frac{R}{i}R'_m \qquad (4\text{-}28)$$

Equation (4-27) is in accord with Eq. (4-23) for the value of an annuity at its origin date. First, assume that an annuity is to consist of a finite number of payments. The value of the annuity at its origin date is

$$_nT'_0 = R\frac{1-(1+i)^{-mn}}{(1+i)^m - 1} \qquad (4\text{-}23)$$

Now, as the number of payments n becomes infinitely large, the term $(1+i)^{-mn}$ approaches zero as a limit. We then have

$$M = \lim_{n\to\infty} {}_nT'_0 = R\frac{1-0}{(1+i)^m - 1}$$

$$= \frac{R}{(1-i)^m - 1} \qquad (4\text{-}27)$$

Example 4-18.   (P.E. examination problem.)   What amount must be donated to build an institution having an initial cost of $500,000, provide an annual upkeep of $50,000, and have $500,000 at the end of each 50-year period to rebuild the institution.   Assume that invested funds return 4 per cent.

Solution.   The annual upkeep of $50,000 can for simplicity be regarded as a single disbursement made at the end of each year.   The required payments can be grouped in the following manner:

1. An initial payment of $500,000 to construct the building.
2. A perpetuity consisting of payments of $500,000 each made at 50-year intervals.
3. A perpetuity consisting of annual payments of $50,000 each for maintenance.

$$M = 500,000 + \frac{500,000}{0.04}R'_{50} + \frac{50,000}{0.04}$$

$$= 500,000 + \left(\frac{500,000}{0.04}\right)0.00655 + 1,250,000$$

$$= \$1,831,875$$

Example 4-19. An endowment fund is to provide an annual scholarship of $4,000 for the first 5 years, $6,000 for the next 5 years, and $9,000 thereafter. The fund will be established 1 year before the first scholarship is awarded. If the fund earns 4½ per cent interest, what sum must be deposited?

Solution. The payments are recorded in Fig. 4-15.

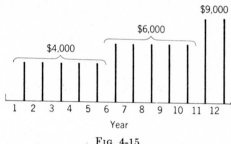

$9,000

$6,000

$4,000

1 2 3 4 5 6 7 8 9 10 11 12

Year

Fig. 4-15

METHOD 1. Selecting the date the fund is established as the valuation date, determine the value of each set of payments.

$$\text{Value of first set} = 4,000 \; {}_5t_0 = \$17,560$$
$$\text{Value of second set} = 6,000 \; {}_5t_0(1.045)^{-5} = \$21,136$$
$$\text{Value of third set} = \frac{9,000}{0.045}(1.045)^{-10} = \$128,786$$
$$M = 17,560 + 21,136 + 128,786 = \$167,482$$

METHOD 2. Assume tentatively that all payments will be $9,000. As a correction, deduct the following sets of savings: an annual saving of $3,000 for the next 10 years and, in addition, an annual saving of $2,000 for the next 5 years. Then

$$M = \frac{9,000}{0.045} - 3,000 \; {}_{10}t_0 - 2,000 \; {}_5t_0$$
$$= 200,000 - 23,738 - 8,780 = \$167,482$$

4-15. Uniform-gradient Series. If payments are made at equal intervals of time and successive payments differ by a constant amount, the series of payments is referred to as a *uniform-gradient* series. Let R denote the first payment and D the constant difference between payments. The second payment is $R + D$, the third payment is $R + 2D$, and the nth payment is $R + (n - 1)D$.

The value of the series of payments at the terminal date (i.e., immediately after the nth payment) is

$$S_n = \left(R + \frac{D}{i}\right) s_{\bar{n}} - n\frac{D}{i} \qquad (4\text{-}29)$$

The value of the series of payments at the origin date (i.e., one period prior to the first payment) is

$$_nT_0 = \left(R + \frac{D}{i} + nD\right)_n t_0 - n\frac{D}{i} \qquad (4\text{-}30)$$

In the special case where $D = R$, Eq. (4-29) reduces to

$$S_n = \frac{R}{i}[s_{\overline{n+1}} - (n+1)] \qquad (4\text{-}29a)$$

In the special case where $D = -R/n$ and the nth payment is therefore R/n, Eq. (4-30) reduces to

$$_nT_0 = \frac{R}{i}\left(1 - \frac{_n t_0}{n}\right) \qquad (4\text{-}30a)$$

Example 4-20. Four end-of-year deposits were made in a sinking fund earning 5 per cent per annum. The first deposit was $3,000 and each deposit thereafter was $400 more than the preceding one. What was the principal in the fund immediately after the last deposit was made?

Solution. The principal can be found by evaluating the individual deposits.

$$S_4 = 3,000(1.05)^3 + 3,400(1.05)^2 + 3,800(1.05) + 4,200$$
$$= 3,472.89 + 3,748.50 + 3,990.00 + 4,200.00 = \$15,411.39$$

Alternatively, the principal can be found by applying Eq. (4-29).

$$R = \$3,000 \qquad D = \$400 \qquad n = 4 \qquad \frac{D}{i} = \frac{400}{0.05} = \$8,000$$

Substituting in the equation,

$$S_4 = 11,000 s_{\bar{4}} - 4 \times 8,000 = 11,000(4.31013) - 4 \times 8,000$$
$$= \$15,411.43$$

Example 4-21. Annual scholarships are to be paid from a fund earning 4 per cent per annum. The first payment will be $1,000. Thereafter, the payments will increase at the rate of $250 each year, making the ninth payment $3,000.

All subsequent payments will be $3,000, with payments to continue indefinitely. What sum must be deposited in the fund 1 year before the first payment is made to provide for this perpetuity?

Solution. METHOD 1. Assume tentatively that all payments will be $3,000. The amount to be deposited is

$$M = \frac{3,000}{0.04} = \$75,000$$

To reflect the true conditions, we must now deduct the value of a series of payments consisting of the following: first payment, $2,000; second payment, $1,750; third payment, $1,500; eighth payment, $250. For this series,

$$R = \$2,000 \qquad n = 8 \qquad \frac{R}{i} = \frac{2,000}{0.04} = \$50,000$$

By Eq. (4-30a),

$$_nT_0 = 50,000\left(1 - \frac{_sl_0}{8}\right) = 50,000\left(1 - \frac{6.73274}{8}\right) = \$7,920$$

Sum to be deposited $= \$75,000 - \$7,920 = \$67,080$

METHOD 2. Select the end of the ninth year as the valuation date. The value of all future payments of $3,000 each is $3,000/0.04$ or $75,000. The value of the first nine payments is found by Eq. (4-29).

$$R = \$1,000 \qquad D = \$250 \qquad n = 9 \qquad \frac{D}{i} = \frac{250}{0.04} = \$6,250$$

$$S_n = 7,250(10.58280) - 9 \times 6,250 = \$20,475$$

Value of all payments $= \$75,000 + \$20,475 = \$95,475$

At the origin date,

$$\text{Value of all payments} = 95,475(1.04)^{-9} = 94,475(0.70259)$$
$$= \$67,080$$

PROBLEMS

4-1. (P.E. examination problem.) A new factory costs $1,000,000. The cost is to be paid off in five equal yearly payments, each payment combining an amortization installment and interest at 4 per cent on the previously unpaid balance of the debt. What should be the amount of each payment? (Carry the answer to five significant figures.)

4-2. If the sum of $3,500 is deposited in a sinking fund at the end of each year for 7 years and the fund earns interest at the rate of 5 per cent per annum, what is the principal in the fund immediately after the seventh deposit is made? *Ans.* $28,497

4-3. In order to retire an issue of bonds whose maturity value is $120,000, a corporation established a sinking fund consisting of 20 equal semiannual deposits, the last deposit falling on the retirement date of the bonds. If the interest rate of the fund is 4 per cent per annum compounded semiannually, what is the amount of each deposit? *Ans.* $4,939

4-4. A loan of $15,000 was to be amortized by a series of 10 uniform semiannual payments, the first payment being made 6 months subsequent to consummation of the loan. Immediately after the fourth payment was made, however, the debtor was in a financial position to discharge the entire debt by a single payment on that date. If the interest rate of the loan was 8 per cent compounded semiannually, what was the amount of this payment? *Ans.* $9,695

4-5. An investor is contemplating the purchase of an annuity that will yield an income of $1,000 at the end of each year for 10 years. If he considers his money to be worth 6 per cent, what is the maximum price the investor should pay for the annuity? *Ans.* $7,360

4-6. A corporation is planning to expand its production facilities in the future. This program entails the following estimated expenditures:

$100,000 at the end of 1980
$100,000 at the end of 1982
$150,000 at the end of 1985

To accumulate the required funds, it established a sinking fund consisting of 15 uniform annual deposits, the first deposit having been made at the end of 1971. The interest rate of the fund is $4\frac{1}{2}$ per cent per annum. Calculate the annual deposit. What will be the balance in the fund on Jan. 1, 1982? *Ans.* $18,703; $154,368

4-7. In order to accumulate the sum of $15,000 in 10 years, a sinking fund is established consisting of 10 annual deposits. If the interest rate is 4 per cent compounded quarterly, what is the amount of the annual deposit? *Ans.* $1,246

4-8. A debt of $30,000 is to be amortized by means of 20 uniform quarterly payments, with interest at 6 per cent compounded quarterly. Calculate the periodic payment if:
a. The first payment is made 3 months after the loan is consummated.
b. The first payment is made 1 year after the loan is consummated.
Ans. $1,748; $1,827

4-9. What sum must be deposited to provide annual scholarships totaling $10,000 if the fund earns interest at the rate of $3\frac{1}{2}$ per cent compounded semiannually?

4-10. (P.E. examination problem.) A steam boiler is purchased on the basis of guaranteed performance. A test indicates that the operating cost will be $300 more per year than guaranteed. If the expected life of the boiler is 20 years and money is worth 6 per cent, what deduction from the purchase price would compensate the purchaser for the extra operating cost? *Ans.* $3,441

4-11. (P.E. examination problem.) A new boiler has just been installed. It is estimated that the maintenance cost will be $30 at the end of each year for the first

10 years and $100 at the end of each year thereafter until the boiler is scrapped at the age of 25 years. What sum of money set aside at this time at 3 per cent will take care of all maintenance expenses for the boiler?   *Ans.* $1,144

4-12. On Jan. 1, 1965, a business firm established a sinking fund for the purchase of a new building at the end of 1974 at an estimated cost of $600,000. The fund was to consist of 10 uniform end-of-year deposits and earned interest at 4 per cent per annum. Owing to financial difficulties, however, deposits were omitted at the end of 1969 and 1970. If the remaining four deposits were equal, what was the amount of these deposits?   *Ans.* $78,060

4-13. (P.E. examination problem.) A yearly expenditure of $300 is required to keep a certain length of roadway pavement in repair. With money at 4 per cent, what expenditure is justified to install a new pavement that may be expected to require no maintenance for the first 10 years, $100 per year for the next 5 years, and $300 per year thereafter?

4-14. (P.E. examination problem.) In order to have funds available to replace equipment purchased on Jan. 1, 1972, a company estimates that it will need $200,000 on the first of January in each of the years 1990 through 1993. Funds are to be built up by investing a fixed sum on Jan. 1, 1973, and yearly thereafter to 1989 inclusive at an annual rate of 3 per cent. What amount must be invested annually?

Ans. $34,160

4-15. An investor paid $2,900 at the beginning of a particular year for an annuity that paid $500 at the end of each year for 7 years. What interest rate did he earn?

4-16. (P.E. examination problem.) One hundred shares of a certain stock were purchased in 1961 for $30 per share. The stock paid dividends of $2 per share for 8 years and then $1.50 per share for 4 years. The stock was then sold for $29 per share. What rate of return was made on the investment?

4-17. An investor purchased a parcel of land for $40,000. The total annual payment for taxes, insurance, and maintenance was $1,700. (This is to be treated as a lump-sum end-of-year payment.) The investor decided to sell the land at the end of the sixth year. Sales commission and legal fees are expected to total 8 per cent of the selling price.

The annual payment associated with this investment served to reduce the investor's regular taxable income, the highest tax rate on his income being 44 per cent. The net profit realized by sale of the land will be taxed as capital gains at 25 per cent. If the investor wishes to earn 8 per cent after taxes, what is the minimum price at which he should offer to sell the land?

Solution. Let S denote the selling price and X the net proceeds from sale of the land.

$$\text{Effective annual payment} = 1,700(1 - 0.44) = \$952$$

Applying an 8 per cent rate,

$$X = 40,000(1.08)^6 + 952s_{\overline{6}} = \$70,459$$
$$\text{Capital-gains tax} = 0.25(0.92S - 40,000) = 0.23S - 10,000$$
$$X = 0.92S - (0.23S - 10,000) = \$70,459$$

Solving,

$$S = \$87,620$$

4-18. A debt of $40,000, whose interest rate is 6 per cent compounded semi-annually, is to be discharged by a series of ten semiannual payments, the first payment to be made 6 months after consummation of the loan. The first six payments will be $3,000 each, while the remaining four payments will all be equal and of such amount that the final payment will liquidate the debt. What is the amount of the last four payments? Verify the solution by constructing an amortization schedule of the loan.

4-19. (P.E. examination problem.) A man, now 20 years of age, will receive a legacy of $50,000 at the age of 30 years. He wishes to study for the next 7 years and estimates he will require $3,500 per year. He will borrow this sum at the beginning of each year and repay it out of his legacy with interest compounded at 6 per cent. How much will be left of the legacy when he receives it?   *Ans.* $12,910

4-20. Annual deposits were made in a fund earning 6 per cent per annum. The first deposit was $1,400 and each deposit thereafter was $90 less than the preceding one. What was the principal in the fund immediately after the seventh deposit was made?   *Ans.* $9,661

4-21. The sum of $20,000 was borrowed on Jan. 1, 1974. The terms for repaying the loan were as follows: The borrower will make five annual payments, the first payment being made on Jan. 1, 1975, and each payment will be $1,000 more than the preceding one. If the interest rate of the loan is 8 per cent, what was the amount of the first payment? Verify the solution by constructing an amortization schedule of the loan.   *Ans.* $3,163

4-22. (This problem is presented for its mathematical interest.) It has been proved in the text that

$$1 + (1 + i) + (1 + i)^2 + \cdots + (1 + i)^{n-1} = \frac{(1 + i)^n - 1}{i}$$

By expanding all binomials, collecting similar terms, and equating, verify the following:

$$1 + 2 + 3 + \cdots + n = \frac{n(n + 1)}{2}$$

$$1 \cdot 2 + 2 \cdot 3 + 3 \cdot 4 + \cdots + n(n + 1) = \frac{n(n + 1)(n + 2)}{3}$$

$$1 \cdot 2 \cdot 3 + 2 \cdot 3 \cdot 4 + 3 \cdot 4 \cdot 5 + \cdots + n(n + 1)(n + 2) = \frac{n(n + 1)(n + 2)(n + 3)}{4}$$

CHAPTER 5

VALUATION OF BONDS

5-1. Introduction. In Chap. 2, we discuss the purposes underlying the issue of bonds and the mechanics involved in the payment of interest and principal. As a review, we present below the basic definitions pertaining to bonds.

The sum of money specified in the bond is termed the face value, or par value, of the bond. The terms of a bond specify the date of maturity of the bond, the interval between interest payments, and the interest rate paid by the issuing organization. This interest rate is to be applied to the face value to obtain the periodic interest payments.

If nothing is stated to the contrary, the bond is to be redeemed at its maturity, with the bondholder receiving the face value of the bond. In many instances, however, the bond carries a provision that at the option of the issuer it can be redeemed during some specified period prior to its maturity. The motives for the insertion of this clause were explained in Chap. 2. Where this redemption feature exists, the bond usually provides that at redemption the bondholder is to receive a sum in excess of the face value of the bond. This sum is generally expressed as a percentage of the face value. For example, if a $1,000 bond carries a provision that it can be redeemed at 105 at some specified date prior to its maturity, then the sum to be paid to the bondholder if this option is exercised is 105 per cent of $1,000, or $1,050. Such a bond is said to be redeemable above par.

In our discussion of bonds, it was emphasized that the price at which a bond is purchased is generally not identical with the face value of the bond. When the purchase price of a bond does coincide with its par value, it is said to be purchased at par; when the purchase price exceeds the par value, it is said to be purchased above par, or at a premium; finally, when the purchase price is below the par value, it is said to be purchased below par, or at a discount. The difference between the par value of a bond and its purchase price is termed the premium or discount, whichever applies. For example, if the face value of a bond is $1,000 and it sells for $940, it has been sold at a discount of $60; if the same bond sells for $1,020, then it has been sold at a premium of $20.

We shall hereafter refer to the periodic interest payments received by the bondholder as the "dividends" and to the interval between successive dividends as the "dividend period."

Our study of the valuation of bonds will be predicated on the same basic assumptions enumerated in Chap. 3. We are assuming that business and financial conditions remain static during the life of the bond, a condition that will be reflected in a stationary investment rate. Therefore, if an investor has purchased a bond at its date of issue at a price to yield a return of 6 per cent on his investment, it follows that at any date prior to the maturity of the bond he can transfer the bond to a new purchaser at such price as to maintain a 6 per cent return on his original investment and to yield the new investor the same rate of return. The value of a bond at any date intermediate between its issue and its redemption is termed its "book value."

Assume that a bond of $1,000 face value, redeemable at par, and earning interest at the rate of 6 per cent annually is purchased by an investor for $980 at a date 1 year prior to its maturity. At the end of the year, the investor receives the following sums of money:

$$\begin{array}{ll}
\text{Dividend, 6 per cent of \$1,000} & = \$\quad 60.00 \\
\text{Redemption price (= face value)} & = \quad 1,000.00 \\
\hline
\text{Total income at end of year} & = \$1,060.00 \\
\text{Investment at beginning of year} & = \quad\ 980.00 \\
\hline
\text{Earning for year} & = \$\quad 80.00 \\
\text{Rate of return on investment} & = {}^{80}\!/_{980} = 8.16 \text{ per cent}
\end{array}$$

Evidently, had the bond been purchased above par, the rate of return realized would have been below the 6 per cent interest rate of the bond. In order to distinguish between these two different rates, we shall refer to the rate of interest quoted in the bond as the "dividend rate" and to the actual rate of return secured on the investment as the "investment rate."

In the above-mentioned numerical example of a bond purchased below par, a brief analysis discloses that there were two factors contributing to the variation of the investment rate from the dividend rate. These are the following:

1. Although the dividend of $60 represents 6 per cent of the face value of the bond, it represents 6.12 per cent of the purchase price, $980.

2. The difference (discount) of $20 between the redemption price of the bond and its purchase price represents a return of 2.04 per cent on the $980 invested.

In a case where a bond is purchased several years prior to its date of maturity, the discount will exert a less pronounced effect upon the investment rate, since this is a monetary gain that is applicable not merely to one year but rather to the number of years remaining to maturity. Moreover, the periodic dividend will represent a variable rate of return on the investment, since the book value of the bond varies from its original purchase price to its maturity value.

Three types of problems usually arise with regard to the valuation of bonds:

1. At what price should a bond be purchased to yield a stipulated investment rate?

2. Conversely, given the purchase price of a bond, what is the corresponding investment rate?

3. In what manner is the premium or discount to be prorated over the life of the investment?

We shall study each of these problems in turn, although not in the order listed above.

5-2. Calculating the Purchase Price of a Bond. To determine the price at which a bond should be purchased to yield a stipulated investment rate, the most obvious method is to evaluate at the date of purchase of the bond and on the basis of the given investment rate all sums of money which the bondholder will receive. This rule applies regardless of whether the bond is to be purchased at its date of issue or at some intermediate date. We shall assume in all cases that the date of purchase coincides with the beginning of a dividend period.

When the bond contains a redemption clause, an element of uncertainty exists regarding the number of dividends to be received, the redemption price, and the redemption date of the bond. For this ambiguous case, the prospective investor should evaluate the bond separately for each alternate set of circumstances and select the lesser value as the intended purchase price.

Example 5-1. A $1,000 bond, redeemable at par in 5 years, pays dividends at the rate of 5 per cent per annum. Calculate the purchase price of the bond and the book value at the beginning of each dividend period, if the bond is purchased to yield a rate of return of (*a*) 5 per cent, (*b*) 8 per cent, and (*c*) 3 per cent.

Solution. Part a. Since the investment rate equals the dividend rate, the purchase price of the bond at its date of issue equals its par value of $1,000. The variation of the value of the bond during its term is depicted in Fig. 5-1, based on the method of chronological sequence. At date of issue, the bond is worth $1,000, represented by *OA*. During the first year, the bond earns interest at the rate of 5 per cent and therefore attains the value of $1,050 at the end of that

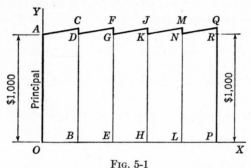

Fig. 5-1

year (*BC*). At this date, however, a dividend of $50 is received by the bond-holder (*CD*), restoring the bond to its original value of $1,000. This process is repeated every period. The result is that at the beginning of each dividend period the value of the bond has reverted to its par value.

Solution. Part b. The sums of money which the bondholder is to receive are the following:

1. The series of five annual dividends, each of which is 5 per cent of the face value, or $50. These dividends constitute an annuity whose value at date of issue of the bond, based on a rate of 8 per cent, is given by Eq. (4-4) as

$$_5T_0 = 50 \left[\frac{1 - (1.08)^{-5}}{0.08} \right] = 50(3.99271) = \$199.64$$

2. The redemption value of the bond, $1,000, which will be received 5 years after the date of purchase. The value of this sum at date of purchase is

$$P_{-5} = 1,000(1.08)^{-5} = 1,000(0.68058) = \$680.58$$

The purchase price of the bond is, therefore,

$$\$199.64 + \$680.58 = \$880.22$$

Hence, the bond is purchased at a discount of $119.78.

The book value of the bond at the commencement of each subsequent year can be calculated by repeating this arithmetical process. The results obtained are recorded in Table 5-1.

TABLE 5-1. BOOK VALUE OF BOND

Beginning of year	Value of future dividends	Value of redemption price	Value of bond
1	$199.64	$ 680.58	$ 880.22
2	165.61	735.03	900.64
3	128.85	793.83	922.68
4	89.16	857.34	946.50
5	46.30	925.93	972.23
End of 5th year	0.00	1,000.00	1,000.00

As we approach the maturity date, there are two opposing forces operating to determine the value of the bond. The first is the decline in the value of the future dividends; the second is the increase in the value of the redemption price. In this instance, where the investment rate exceeds the dividend rate, the second

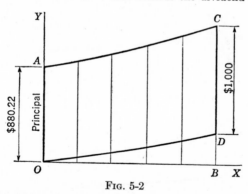

FIG. 5-2

force outweighs the first in significance. The result is a continual increase in the value of the bond, from its original purchase price of $880.22 to its redemption value of $1,000.

In calculating the bond value at an intermediate date, we can modify the above procedure by evaluating at that date all sums of money pertaining to the past rather than the future. This method requires evaluating at the date in question the original purchase price of the bond and then deducting from it the value of the dividends already received. Figure 5-2 is a graphical representation of the variation of book value, based on this method of calculation. OA represents the original purchase price, which expands in value along curve AC. Curve OD is the annuity-principal curve, representing the value of the dividends received up to a given date. The value of the bond at any date therefore equals the difference between the ordinates of the two curves. Hence, the ordinate DC at the maturity date equals the redemption price of the bond.

Let D denote the periodic dividend and i the investment rate at which the bond is purchased. In accordance with Eq. (4-3), the value of the dividends received up to a given date equals the interest earned by the sum D/i, if invested at the origin date of the annuity. If we refer to Fig. 5-2, it is seen that when the purchase price is less than D/i, the two curves converge as we proceed to the right, which means that the purchase price exceeds the redemption price. The reverse is true if the purchase price is greater than D/i. On the other hand, if the purchase price equals D/i, then the two curves are parallel and the purchase price equals the redemption price.

The results of this analysis suggest an alternate method of calculating the purchase price of a bond, based on the factor D/i. In the problem at hand, assume for convenience that the annual dividend of $50 actually did represent a return of 8 per cent on the investment. If this were true, both the purchase price and the redemption price of the bond would be $50/0.08, or $625. Since the true redemption price is $1,000, the purchase price must be augmented by the difference of $375 between the assumed and the true maturity value, this difference, however, being translated to its equivalent value at the date of purchase. Applying this method, we have

$$\begin{aligned}
\text{Purchase price} &= 625 + 375(1.08)^{-5} \\
&= 625 + 375(0.68058) \\
&= 625 + 255.22 \\
&= \$880.22
\end{aligned}$$

From this point of view, the purchase of the bond for $880.22 can be conceived as being composed of two distinct loans extended by the buyer to the issuer, one for $625 and the other for $255.22. Both loans earn interest at the rate of 8 per cent. The interest earned by the first loan is paid at the end of each year, while the interest earned by the second loan is compounded. Thus, the interest-earning principal remains constant for the first loan but varies for the second. The value of the bond at any intermediate date equals the total of the principal of the two loans at that date. At maturity, both loans are liquidated by a payment of the entire principal.

The variation of bond value based on this conception is represented graphically in Fig. 5-3. OA represents the principal of the first loan ($625). Because this principal reverts to its original value at the end of each year, we shall represent this principal during the term of the bond by the straight line AD. Line CE represents the maturity value of the bond ($1,000), leaving a value of $375 for DE. With line AD as our axis for measuring ordinates and with an interest rate of 8 per cent, a principal-time curve is drawn through point E, intersecting the YY axis at B. Line AB therefore represents the original amount of the second loan ($255.22), and OB represents the purchase price of the bond ($880.22).

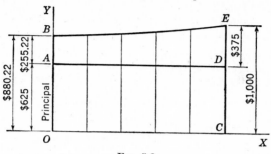

FIG. 5-3

Curve BE is therefore the graph of the bond value at all dates from date of purchase to date of maturity.

During any given year,

Interest earned by bond = interest earned by first loan + interest earned by second loan

Dividend received = interest earned by first loan

∴ Increase in value of bond = interest earned by second loan

Applying this relationship, the variation of bond value is shown in Table 5-2. This table records the growth of the principal of the second loan, based on an

TABLE 5-2. BOOK VALUE OF BOND BY DUAL-LOAN METHOD

Year	Principal of second loan at beginning	Interest earned by second loan	Principal of second loan at end	Value of bond at end
0	$ 0.00	$ 0.00	$255.22	$ 880.22
1	255.22	20.42	275.64	900.64
2	275.64	22.05	297.69	922.69
3	297.69	23.82	321.51	946.51
4	321.51	25.72	347.23	972.23
5	347.23	27.78	375.01	1,000.01

interest rate of 8 per cent. The bond value is then obtained by adding this principal to the principal of the first loan, which remains $625 at the end of each year. The advantage inherent in this method of determining the bond value is that it requires the calculation of only one variable quantity, namely, the principal of the second loan.

The total interest earned by the second loan is $119.79, which is also the discount at which the bond was purchased.

Finally, the variation of the value of the bond during its life can also be calculated by applying the method of chronological sequence, but treating the purchase of the bond as one integral investment. This method is represented graphically in Fig. 5-4. At its date of purchase, the bond is worth $880.22, represented by OA. At the end of the first year, interest has been earned at the rate

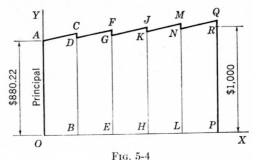

Fig. 5-4

of 8 per cent of the purchase price, or $70.42. Hence, the value of the bond at the end of the first year, but immediately preceding the receipt of the first dividend, is $880.22 + 70.42, or $950.64 ($BC$). On that date, a $50 dividend is received (CD), and the value of the bond is reduced to $900.64 ($BD$). Since the interest earning exceeds the annual dividend, the bond value at the commencement of the second year exceeds the purchase price; this increase in value is accelerated with the passage of time.

Similarly, during the second year the bond earns interest at the rate of 8 per cent of its book value at the beginning of that year. This interest amounts to $72.05; the dividend received at the end of the year is again $50, thereby producing at the beginning of the third year a bond value of $922.69 ($EG$). This process is repeated each year until at the end of the fifth year the bond attains its maturity value of $1,000.

Table 5-3 records the variation of the value of the bond, based on the method of chronological sequence.

The situation analyzed above of a bond purchased at a discount can be viewed in another manner. In purchasing a bond, the investor has extended a loan to

TABLE 5-3. BOOK VALUE OF BOND

Year	Bond value at beginning	Interest earned	Dividend received	Bond value at end
1	$880.22	$70.42	$50.00	$ 900.64
2	900.64	72.05	50.00	922.69
3	922.69	73.82	50.00	946.51
4	946.51	75.72	50.00	972.23
5	972.23	77.78	50.00	1,000.01

Total interest earned = $369.79
Total dividends received = 250.00
Total increase in bond value (= discount of bond) = $119.79

the issuer, the principal of the loan being the purchase price of the bond. This loan is earning interest at the rate of 8 per cent per annum. At the end of each period, a portion of the interest earned ($50 in this case) is paid to the creditor by the debtor, while the balance of the interest earned is converted to principal (i.e., compounded). For example, at the end of the first year, the loan has earned interest of $70.42. Of this amount, $50 is paid directly to the bondholder, and the residue of $20.42 is added to the principal of the loan. Hence, a loan of this nature can be regarded as lying intermediate between the two most disparate types of loans: one in which the interest is paid in full to the creditor at the end of each period; the other in which the entire interest earning is converted to principal.

Solution. Part c. The methods of solution developed above are likewise applicable to this problem, with all calculations based on an interest rate of 3 per cent. Applying the first method, we evaluate all future sums of money to be received by the bondholder, including both the annual dividends of $50 each and the redemption price of $1,000. Thus, the purchase price is

$$50 \left[\frac{1 - (1.03)^{-5}}{0.03} \right] + 1,000(1.03)^{-5} = \$228.99 + \$862.61 = \$1,091.60$$

The bond is therefore purchased at a premium of $91.60. The value of the bond at any intermediate date is recorded in Table 5-4.

In this case, as we approach the date of maturity, the decline in the value of the future dividends is of greater consequence than the rise in the redemption value. This produces a gradual contraction in the bond value until at the end of the fifth year it attains its redemption price of $1,000.

We shall now apply the method of solution in which the periodic dividend is equated with the interest earning. If the annual dividend of $50 actually repre-

TABLE 5-4. BOOK VALUE OF BOND

Beginning of year	Value of future dividends	Value of redemption price	Value of bond
1	$228.99	$ 862.61	$1,091.60
2	185.85	888.49	1,074.34
3	141.43	915.14	1,056.57
4	95.67	942.60	1,038.27
5	48.54	970.87	1,019.41
End of 5th year	0.00	1,000.00	1,000.00

sented a return of 3 per cent on the investment, then both the purchase price and the redemption price would be $50/0.03, or $1,666.67. Since, however, the redemption price is only $1,000, the purchase price must be diminished by the difference of $666.67 translated to its equivalent value at date of purchase. We thus have

$$\text{Purchase price} = 1{,}666.67 - 666.67(1.03)^{-5}$$
$$= 1{,}666.67 - 575.07$$
$$= \$1{,}091.60$$

From this point of view, the purchase of the bond for $1,091.60 can be conceived as representing the combined effect of two distinct loans—one for $1,666.67 extended by the buyer to the issuer, the other for $575.07 extended by the issuer to the buyer. Both loans earn interest at the rate of 3 per cent. The interest earned by the first loan is paid at the end of each year, while the interest earned by the second loan is compounded. Thus, the interest-earning principal remains constant for the first loan but varies for the second. The value of the bond at any intermediate date equals the algebraic total of the principal of the two loans at that date. At maturity, both loans are discharged by a payment of the entire principal.

The variation of bond value based on this conception is represented graphically in Fig. 5-5. OA represents the principal of the first loan, $1,666.67. Since this principal reverts to its original value at the end of each year, we shall represent this principal during the life of the bond by the straight line AD. Line CE represents the maturity value of the bond ($1,000), leaving a value of $666.67 for DE. With line AD as our axis for measuring ordinates and with an interest rate of 3 per cent, an inverted principal-time curve is drawn through point E, intersecting the YY axis at B. Line AB therefore represents the original amount of the second loan ($575.07), and OB represents the purchase price of the bond

($1,091.60). Curve *BE* is therefore the graph of the bond value at all dates from date of purchase to date of maturity. During any given year,

Interest earned by bond = interest earned by first loan

            − interest earned by second loan

Dividend received = interest earned by first loan

∴ Decrease in value of bond = interest earned by second loan

From the application of this relationship, the variation of bond value is shown in Table 5-5. This table records the variation of the principal of the second loan,

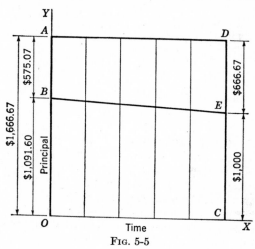

Fig. 5-5

based on an interest rate of 3 per cent. The bond value is then obtained by deducting this principal from the principal of the first loan, which remains $1,666.67 at the end of each year.

TABLE 5-5. BOOK VALUE OF BOND BY DUAL-LOAN METHOD

Year	Principal of second loan at beginning	Interest earned by second loan	Principal of second loan at end	Value of bond at end
0	$ 0.00	$ 0.00	$575.07	$1,091.60
1	575.07	17.25	592.32	1,074.35
2	592.32	17.77	610.09	1,056.58
3	610.09	18.30	628.39	1,038.28
4	628.39	18.85	647.24	1,019.43
5	647.24	19.42	666.66	1,000.01

The total interest earned by the second loan is $91.59, which is also the premium at which the bond was purchased.

Finally, we shall calculate the variation of the value of the bond during its life by applying the method of chronological sequence, as represented graphically in Fig. 5-6. At its date of issue, the bond is worth $1,091.60, represented by *OA*.

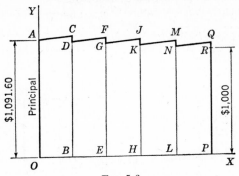

Fig. 5-6

At the end of the first year, interest earned at the rate of 3 per cent of the purchase price amounts to $32.75. Hence, the value of the bond at the end of the first year, but immediately preceding the receipt of the first dividend, is $1,124.35 (*BC*). On that date, a $50 dividend is received (*CD*), and the value of the bond is thereby depressed to $1,074.35 (*BD*). Since the annual dividend exceeds the interest earning, the bond value at the commencement of the second year is less than the purchase price; this decrease in value is accelerated with the passage of time.

Table 5-6 records the variation of the value of the bond, based on the method of chronological sequence.

TABLE 5-6. BOOK VALUE OF BOND

Period	Bond value at beginning	Interest earned	Dividend received	Bond value at end
1	$1,091.60	$32.75	$50.00	$1,074.35
2	1,074.35	32.23	50.00	1,056.58
3	1,056.58	31.70	50.00	1,038.28
4	1,038.28	31.15	50.00	1,019.43
5	1,019.43	30.58	50.00	1,000.01

Total dividends received = $250.00
Total interest earned = 158.41
Total decrease in bond value (= premium of bond) = $ 91.59

Where a bond was purchased at a discount, we found that we can **regard** each dividend as a partial payment of the interest earned during that period, the remainder of the interest earning being converted to principal. In the case of a bond purchased at a premium, on the other hand, the periodic dividend exceeds the interest earning. For this situation, we can consider each dividend to consist of two elements. The first is a payment by the debtor to the creditor of the interest earned during the period, while the second element is a partial payment of principal.

Thus, in the numerical example above, the original principal of the loan is $1,091.60. At the end of the first year, the bondholder has earned interest of $32.75. When he receives his dividend of $50, he is in effect receiving the interest he has earned, with the excess of $17.25 applied to reduce the principal of the loan. Hence, this type of loan is similar to an amortized mortgage. In the latter case, the principal is steadily reduced until at the maturity of the loan it vanishes entirely; in the former case, the principal is steadily reduced until at maturity it attains the redemption price of the bond.

5-3. Bond-value Equations. We shall now derive a general equation pertaining to the value of a bond. Let

B = value of the bond at its date of purchase
D = periodic dividend
i = intended investment rate
n = number of dividends to be received
M = redemption price

Setting the purchase price equal to the total value of the future dividends and the redemption price, we obtain

$$B = D \frac{1 - (1 + i)^{-n}}{i} + M(1 + i)^{-n} \tag{5-1}$$

The bond can also be evaluated in the manner developed in the numerical problem above, in which we first assume the periodic dividend to represent the intended rate of return i on the investment. If this condition were true, the purchase price and redemption price would both be D/i. Since the true redemption price is M, we must correct the purchase price for the difference between the true and hypothetical redemption price, this difference being evaluated at date of purchase.   This procedure yields

$$B = \frac{D}{i} + \left(M - \frac{D}{i} \right)(1 + i)^{-n} \qquad (5\text{-}2)$$

This result can also be obtained from Eq. (5-1) by a slight transfer of terms.

5-4. Amortization of Premium and Accumulation of Discount.   When bonds are sold at a premium or a discount, it is necessary to maintain the accounting records of the issuing organization in such manner as to reflect the true interest expense for each period, as distinguished from the dividends paid, and the true value of the bonds, as distinguished from their face value.   To study the mechanics involved in this procedure, we shall employ the numerical problems solved above.

On the date bonds are sold, an entry is made on the books of the issuing corporation, recording the proceeds obtained from the sale, establishing a liability account called Bonds Payable for the face value of the bonds sold, and establishing a third account called Premium (or Discount) on Bonds Payable, whichever applies.   Thus, assume that a corporation sells 100 bonds of $1,000 face value each, maturing in 5 years and bearing interest at 5 per cent per annum, for a price of $880.22 each.   This corresponds to an investment rate of 8 per cent.   The following accounting entry records this transaction:

A charge to Cash for $88,022
A charge to Discount on Bonds Payable for $11,978
A credit to Bonds Payable for $100,000

The account Discount on Bonds Payable is an adjustment account, since it is to be deducted from Bonds Payable to reflect the true value of the bonds sold.

At the end of the first year, a dividend payment of $5,000 is made, but the true interest expense for the issuing corporation is, according to Table 5-3 above, $7,042.   The accounting entry to record this dividend payment therefore consists of:

A charge to Interest Expense for $7,042
A credit to Cash for $5,000
A credit to Discount on Bonds Payable for $2,042

The last account has thus been reduced from its original value of $11,978 to a new balance of $9,936. Hence, at the end of the first year, the true value of the bonds issued is determined as follows:

$$\begin{array}{rl}
\text{Bonds Payable} & = \$100,000 \\
\text{Discount on Bonds Payable} & = 9,936 \\
\hline
\text{Value of bonds outstanding} & = \$90,064 \\
\hline
\end{array}$$

The total bond discount, which is the difference between the purchase price and redemption price, equals the difference between the total interest earnings and the total dividend payments. Consequently, by the end of the fifth year, the account Discount on Bonds Payable will vanish entirely as a result of the annual reductions made in this account.

We shall now consider the case of bonds sold at a premium. Assume that the 100 bonds described above were sold at a price of $1,091.60 each, which corresponds to an investment rate of 3 per cent. The entry to record the sale of the bonds consists of:

A charge to Cash for $109,160
A credit to Bonds Payable for $100,000
A credit to Premium on Bonds Payable for $9,160

Thus, the true value of the bonds outstanding is obtained by totaling the balances in the accounts Bonds Payable and Premium on Bonds Payable.

If we refer to Table 5-6 above, it is seen that at the end of the first year the interest expense of the issuing corporation is $3,275, while the dividends paid are $5,000. Therefore, the entry to record the payment of the dividend consists of:

A charge to Interest Expense for $3,275
A charge to Premium on Bonds Payable for $1,725
A credit to Cash for $5,000

The balance in the account Premium on Bonds Payable has thus been reduced from its original value of $9,160 to $7,435. Hence, at the end of the first year, we obtain the true value of the bonds sold in the following manner:

$$\begin{array}{rl}
\text{Bonds Payable} & = \$100,000 \\
\text{Premium on Bonds Payable} & = 7,435 \\
\hline
\text{Value of bonds outstanding} & = \$107,435 \\
\hline
\end{array}$$

As noted above, the account Premium on Bonds Payable will vanish by the end of the fifth year.

The process outlined above of systematically extinguishing the account Premium (or Discount) on Bonds Payable by charging or crediting it periodically for the disparity between interest expense and dividend payment is known as "amortizing the bond premium," or "accumulating the bond discount." In order to accomplish this in the exact manner shown, it is necessary to construct a table recording the variation of the value of a bond during its life. In actual practice, this is not frequently done. Instead, an average correction is obtained by dividing the total premium or discount by the number of dividend periods included in the life of the bond. The account Premium (or Discount) on Bonds Payable is then charged or credited each period for the amount thus obtained. For example, in the case of the bonds described above, whose par value is $100,000 and selling price $88,022, the annual credit to Discount on Bonds Payable will be $11,978/5, or $2,395.60. This is known as the "straight-line method" of amortizing the premium or accumulating the discount.

5-5. Total Periodic Disbursement of Debtor.   Assume that, in order to accumulate the funds necessary to retire an issue of bonds at its maturity, the seller will make periodic deposits in a sinking fund, the deposit period of the fund coinciding with the dividend period of the bonds.   Let

F = total face value of the bonds
M = total redemption value of the bonds
d = dividend rate of the bonds
i = interest rate of sinking fund

The total disbursement pertaining to the bonds which the seller must make at the end of each period consists of the following:

1. The periodic dividend D payable to the bondholders

$$D = Fd$$

2. The periodic contribution R to the sinking fund

$$R = M \frac{i}{(1 + i)^n - 1} = MR'_n$$

3. The administrative expenses incidental to the periodic payment of dividends

In many numerical problems, the administrative and legal expenses associated with the bonds are excluded from consideration, even though they constitute a substantial element of cost. We shall therefore disregard these expenses in the material that follows. We then have

$$\text{Total periodic disbursement} = Fd + MR'_n \qquad (5\text{-}3)$$

Moreover, if the bonds are sold at par and are redeemable at par ($M = F$), then this periodic payment will represent the actual periodic cost of the bonds. If, in addition, the interest rate i of the sinking fund and the dividend (or investment) rate d of the bonds are equal, we obtain

$$\begin{aligned} \text{Periodic cost of bonds} &= \text{total periodic disbursement} \\ &= Fi + FR'_n = F(i + R'_n) \end{aligned}$$

By Eq. (4-6),

$$R''_n = i + R'_n$$

Applying this relationship to the last formula yields

Periodic cost of bonds $= FR''_n$

$\qquad\qquad$ = periodic payment of annuity whose date-of-origin value is F

for the special case where the bonds are sold at par, are redeemable at par, and have an investment rate equal to the interest rate of the sinking fund.

Example 5-2. (P.E. examination problem.) A community wishes to purchase an existing utility valued at $500,000 by selling 5 per cent bonds that will mature in 30 years. The money to retire the bonds will be raised by paying equal annual amounts into a sinking fund that will earn 4 per cent. What will be the total annual cost of the bonds until they mature?

Solution

$$F = M = \$500{,}000 \qquad d = 5 \text{ per cent} \qquad i = 4 \text{ per cent} \qquad n = 30$$
$$R'_{30} \text{ for 4 per cent rate} = 0.01783$$
$$\begin{aligned} \text{Annual cost} &= 500{,}000(0.05) + 500{,}000(0.01783) \\ &= 25{,}000 + 8{,}915 \\ &= \$33{,}915 \end{aligned}$$

Implicit in this solution is the assumption that the bonds can actually be sold at par, although such is not necessarily the case.

Example 5-3. (P.E. examination problem.) A municipality wishes to raise funds for improvements by issuing 5½ per cent bonds. There is $20,000 available per year for interest payments and retirement of the bonds at 110. What should be the amount of the bond issue if all the bonds are to be retired in 20 years?

Solution. The problem omits mention of the interest rate of the sinking fund; we shall assume this to be 3 per cent.

$$M = 1.10F \qquad d = 5\tfrac{1}{2} \text{ per cent} \qquad i = 3 \text{ per cent} \qquad n = 20$$

$$\text{Annual disbursement} = \$20,000$$

$$R'_{20} \text{ for 3 per cent rate} = 0.03722$$

Then

$$20,000 = 0.055F + 1.10F(0.03722)$$

Solving for F,

$$F = \$208,400$$

5-6. Calculation of Interest Rate of Bond. In the preceding material, we calculated the purchase price of a bond corresponding to a particular interest rate. In many cases, however, the purchase price of a bond is the known quantity, and it is necessary to determine the investment rate secured by the purchaser. A calculation of this nature is also required to enable the issuing corporation to determine the true rate of interest it is paying for borrowed money. The basis for calculating this interest rate is not the actual selling price of the bonds but rather the sum of money remaining after the payment of administrative and legal expenses.

Assuming there are no bond tables available for obtaining the interest rate corresponding to a given purchase price, we shall calculate the approximate rate in the manner presented in Art. 3-11.

Example 5-4. (P.E. examination problem.) A man paid $1,100 for a $1,000 bond that pays $40 per year. In 20 years the bond will be redeemed for $1,050. What net rate of interest will the man obtain on his investment?

Solution. Let i denote the interest rate. The controlling relationship is

$$\text{Value of money paid} = \text{value of money received}$$

Evaluating all sums at the maturity date, we obtain

$$1100(1 + i)^{20} = 40\frac{(1 + i)^{20} - 1}{i} + 1050$$

To obtain a first approximation, we substitute $(1 + ni)$ for the expression $(1 + i)^n$. Then

$$1100(1 + 20i) = 40(20) + 1050$$

$$i = 3.41 \text{ per cent} \qquad \text{as a first approximation}$$

Since we have understated the value both of the \$1,100 investment and of the periodic dividends, it is reasonable to presume that this value does not deviate markedly from the true value. Try a $3\frac{1}{2}$ per cent rate:

$$1100(1.035)^{20} = 1100(1.98979) = \$2,188.77$$
$$40s_{\overline{20}} = 40(28.27968) = \underline{1,131.19}$$
$$\text{Redemption price for } 3\frac{1}{2} \text{ per cent rate} = \underline{\$1,057.58}$$

Try a 3 per cent rate:

$$1100(1.03)^{20} = 1100(1.80611) = \$1,986.72$$
$$40s_{\overline{20}} = 40(26.87037) = \underline{1,074.82}$$
$$\text{Redemption price for 3 per cent rate} = \underline{\$\ \ 911.90}$$

Applying straight-line interpolation, we obtain

$$i = 3.47 \text{ per cent} \qquad \text{approximately}$$

PROBLEMS

5-1. To secure a return of 4 per cent, at what price should a bond be purchased if it is redeemable at \$1,000 in 10 years and pays annual dividends of \$35?

Ans. \$959.44

5-2. What is the purchase price of a \$5,000 bond, redeemable at par in 8 years and paying interest annually at 4 per cent, in order to yield a $4\frac{1}{2}$ per cent return to the investor? Construct a schedule of the accumulation of discount, using the exact method.

5-3. A corporation issued 100 sinking-fund bonds of \$1,000 face value, redeemable at par in 15 years, with interest payable semiannually at 4 per cent per annum. The sinking fund earns interest at the rate of 3 per cent compounded semiannually, and contributions to the fund are made twice a year. Calculate the total periodic disbursement of the corporation for these bonds, exclusive of legal and administrative expenses.

Ans. \$4,664

5-4. To finance a public works program, a municipality will issue $3\frac{1}{2}$ per cent sinking-fund bonds, redeemable at par in 15 years, with interest payable annually. The sinking-fund rate is 3 per cent per annum. If it is estimated that \$25,000 can be raised annually through taxation to provide for the interest payment and sinking-fund deposit, what is the maximum face value of the total bond issue? Assume the bonds are sold at par. *Ans.* \$281,600 (to the nearest one hundred dollars)

5-5. An investor paid \$4,900 to purchase a \$5,000 bond, redeemable at 105 in 6 years and paying interest annually at 4 per cent. What is the investment rate corresponding to this purchase price? As a means of verifying the solution, construct a schedule of the accumulation of discount, using the exact method.

5-6. (P.E. examination problem.) A \$1,000,000 issue of 3 per cent 15-year bonds was sold at 95. If miscellaneous initial expenses of the financing were \$20,000 and a yearly expense of $2,000 is incurred, what is the true rate that the company is paying for the money it borrowed?   *Ans.* 3.82 per cent

5-7. An investor purchased a $1,000 bond for $940. After he received eight annual dividends of $45 each, he was granted an option of converting the bond to 10 shares of stock of the issuing corporation. If the desired investment rate is 6 per cent and the investor plans to sell the stock immediately after its acquisition, what is the minimum market value of the stock that will warrant exercising the option?

5-8. A $1,000 bond, redeemable at par in 20 years and paying dividends of $40 annually, was purchased at a price to yield a 5½ per cent investment rate. What was the true interest earning during the thirteenth year?   *Ans.* $49.77

5-9. A $5,000, twenty-year bond is redeemable in installments in the following manner:

   $700 at the expiration of 5 years
    700 at the expiration of 10 years
    700 at the expiration of 15 years
   $2,900 at the expiration of 20 years

Interest is payable annually on the unredeemed balance at the rate of 6 per cent. What is the purchase price corresponding to an investment rate of 5 per cent? (A bond of this nature is termed an "installment bond.")   *Ans.* $5,518.43

5-10. In the preceding problem, assume that 100 bonds of $5,000 par value have been issued. A sinking fund is established at the date of issue of the bonds to provide for their redemption. Compute the semiannual deposit required if the interest rate of the fund is 4 per cent compounded semiannually. To verify the solution, evaluate all sums of money at both the date of issue and date of maturity.

5-11. A corporation floated an issue of 15-year bonds to finance an expansion program. Anticipating that the earnings of the firm would increase in later years as the effects of this program asserted themselves, it provided that interest on the bonds was to be paid annually at the rate of 4 per cent for the first 10 years and at the rate of 5 per cent for the remaining 5 years. An investor purchased a $1,000 bond at a price to yield 5½ per cent. What was the purchase price?

5-12. A $1,000, ten-year bond earns interest annually at 3 per cent for the first 5 years and 4½ per cent for the remaining 5 years. Calculate the purchase price of the bond to yield an investment rate of 4 per cent if the bond is purchased

   a. At its date of issue.
   b. Five years prior to maturity.   *Ans.* $973.77; $1,022.26

It is interesting to observe the manner in which the value of a bond varies when the dividends are not uniform. In this particular case, the interest earnings exceed the dividends during the first 5 years of the term of the bond, and the bond value therefore increases, surpassing the par value at the expiration of 3 years. After the fifth year, however, a reversal occurs, and the excess of dividend over interest earning causes the bond value to recede, diminishing to the par value at date of maturity.

CHAPTER 6

DEPRECIATION AND DEPLETION

6-1. Introduction. The monetary value of every physical asset varies with the passage of time. There are many factors of an external nature, such as economic and technological changes, that affect this value. In addition to these, however, there is a process in continuous operation that acts inexorably to reduce the value inherent in the asset. This deterioration process is the "wearing out" of an asset due to its use and the action of the elements upon it, and it is referred to as "depreciation." Proper care and maintenance can retard the process of depreciation, but they cannot arrest it.

As a result of depreciation, a point is eventually reached at which it becomes economical to replace an existing machine, building, or other asset with a new one. The exact determination of this economical point of replacement is, of course, a matter of personal judgment, but it is inevitable that every asset used commercially will require replacement.

When an existing asset is to be replaced, it may still retain a certain monetary value. For example, in the case of a machine, some components may still be in good operating condition, the metal itself may possess value as scrap, etc. From this residual value, however, must be deducted the cost of dismantling and transportation. The remaining monetary value, which represents the net proceeds obtainable from sale of the asset, is termed the "salvage value."

The difference between the price at which an asset is purchased and its salvage value when it is eventually retired is the total depreciation which the asset undergoes, and it is referred to as the "wearing value" of the asset. This total depreciation represents a loss of the capital invested, and must be prorated over the life of the asset. However, the salvage value of the asset and the extent of its service life remain unknown quantities until the asset is actually retired. Consequently, in order to allow

113

for depreciation while the asset is still in operation, it is necessary to estimate both the salvage value and the longevity of the asset. In many cases, these estimates are not left to the discretion of the business firm owning the asset, for the Federal government prescribes definite rules concerning depreciation for the purpose of determining taxable income.

The estimates made regarding the salvage value and longevity of an asset may prove, of course, to be highly erroneous. Thus, a machine that has been abused by its operators or that was improperly constructed may require replacement far sooner than anticipated. The reverse conditions are also possible. Corrections can be made in the accounting procedure as it becomes evident that the original estimates were substantially in error.

Assuming that a reasonably correct estimate has been made for the longevity and salvage value of a given asset, we are next confronted with the mathematical question: In what manner does the value of the asset vary from its original value at date of purchase to its salvage value at date of retirement? To rephrase the question, in what manner shall the total depreciation that occurs during the life of the asset be prorated among the various years of its life? This is a question that cannot be answered with mathematical precision, and any reasonable assumption is valid. It appears that depreciation is maximum during the first year of an asset's life and gradually tapers off thereafter. In other words, the process of depreciation is decelerated with the passage of time.

The net value of an asset at an intermediate date, which equals the original cost less the depreciation that has occurred up to that date, is referred to as its "book value." Fig. 6-1 represents the variation in the book value of the asset during its life, based on the general pattern of depreciation which we are assuming to be true. The depreciation for a particular year is the difference between the ordinates at the beginning and end of that year. The rate of depreciation at any given date is equal to the absolute value of the slope of the curve at the corresponding point.

This presumed book-value curve can be used as a criterion for evaluating each method of allocating depreciation by determining the degree to which the results of the method conform to this curve.

We shall assume that the books of the concern owning the asset are closed annually. The accounting entry to record the annual depreciation consists of a charge to the account Depreciation for the amount of depreciation computed for that year and a credit to the account Reserve for

Depreciation for the same amount. Hence, the accumulated balance in the latter account at any intermediate date represents the amount of depreciation the asset has undergone up to that date. The Reserve for Depreciation is an adjustment account; its balance is to be deducted from the original cost of the asset to obtain the value of the asset at this intermediate date.

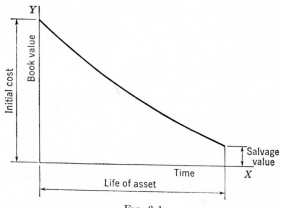

FIG. 6-1

The following methods of apportioning depreciation are the ones most frequently applied:

1. The straight-line method
2. The sinking-fund method
3. The fixed-percentage (declining-balance) method
4. The double-declining-balance method
5. The sum-of-integers method

6-2. Straight-line Method. This is the simplest method of all and is therefore the one most extensively employed. By this method, the total depreciation (wearing value) is allocated uniformly to each year of the asset's life. Thus, assume a machine costs $15,000 and has an estimated salvage value of $3,000 and an estimated life of 8 years. The total depreciation is $12,000; dividing this by 8 years gives an annual depreciation charge of $1,500. In general, let

C_0 = original cost of the asset
L = salvage value
W = wearing value ($= C_0 - L$)

D = annual depreciation charge

n = life expectancy of asset, in years

By the straight-line method,

$$D = \frac{W}{n} \tag{6-1}$$

Let C_r denote the book value of the asset at the end of the rth year. Since the depreciation charged up to that date is rD, we have

$$C_r = C_0 - rD = C_0 - r\frac{W}{n} \tag{6-2}$$

Hence, if a graph of book value is constructed with time as the abscissa, the graph is a straight line; this property explains the nomenclature.

Although the straight-line method of charging depreciation is often preferred because of its innate simplicity, it is somewhat inaccurate by virtue of the fact that it assumes a uniform rate of depreciation, an expedient which is untrue. The book values obtained by this method are therefore inflated to a varying degree.

6-3. Sinking-fund Method. Our study of finance is based upon the assumption that the business and technological conditions existing at a particular instant remain permanent. As a corollary of this assumption, it follows that at the date of retirement of an asset it will be possible to replace this asset with a new one whose purchase price is identical with the original one. Consequently, the sum of money required to make this replacement will be the cost C_0 of the new asset less the salvage value L of the old asset, the difference between the two being the wearing value W. The business firm owning the asset can assure itself of having this sum available by making periodic contributions to a sinking fund that earns a stipulated rate of interest, the periodic deposit being of such amount that at the date of retirement the principal in the fund will total W.

Assume for convenience that the date of deposit of the fund is also the date for closing the books of the business firm. Now, since the sinking-fund principal at date of retirement equals the total depreciation W of the asset, we can likewise equate the principal in the fund at any intermediate date with the total depreciation occurring up to that date. Hence, we consider the asset to be depreciating at the same rate at which the sinking-fund principal is developing. During any year, the principal in the fund increases through the interest earned by the fund and the deposit made at the end of the year.

Consider that a sinking fund is established as a means of financing the replacement of a particular asset; assume that the annual deposit in the fund is $500 and the interest earned by the fund during one particular year is $30.45. The accounting entry to record the depreciation for the year will then consist of:

A charge to Depreciation for $530.45
A credit to Cash for $500.00
A credit to Interest Income for $30.45

In reality, since a prospering business firm can usually secure a greater return from investing capital in its own enterprise rather than in a fund, the sinking fund whose existence we have assumed for theoretical reasons is not actually created but exists solely in the imagination. It is, so to speak, a measuring rod, revealing the amount of depreciation to be charged each year. It therefore becomes necessary to assume an interest rate for this hypothetical fund.

In the numerical illustration above, if no sinking fund has been created, the actual entry to record depreciation for that particular year will be:

A charge to Depreciation for $530.45
A credit to Reserve for Depreciation for $530.45

The straight-line method of calculating depreciation may be regarded as a special case of the sinking-fund method, in which the interest rate is zero.

Example 6-1. An asset costing $20,000 has a life expectancy of 5 years and an estimated salvage value of $3,000. Employing the sinking-fund method, calculate the annual depreciation and end-of-year book value of the asset. Use an interest rate of 4 per cent.

Solution

$$S_n = \$20,000 - \$3,000 = \$17,000$$
$$n = 5 \qquad i = 4 \text{ per cent}$$
$$R = 17,000R'_5 = 17,000(0.18463)$$
$$= \$3,138.71$$

The information required is recorded in Table 6-1.

TABLE 6-1. DEPRECIATION SCHEDULE FOR SINKING-FUND METHOD

Year	Principal in fund at beginning	Interest earned	Depreciation charge	Balance in reserve at end	Book value at end
0	. . . .	. . . .	. . . .	. . . .	$20,000.00
1	. . . .	. . . .	$3,138.71	$ 3,138.71	16,861.29
2	$ 3,138.71	$125.55	3,264.26	6,402.97	13,597.03
3	6,402.97	256.12	3,394.83	9,797.80	10,202.20
4	9,797.80	391.91	3,530.62	13,328.42	6,671.58
5	13,328.42	533.14	3,671.85	17,000.27	2,999.73

The annual depreciation is obtained by adding the interest earned by the (imaginary) fund for that year to the constant annual deposit of $3,138.71.

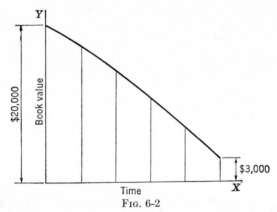

FIG. 6-2

The decline in book value as based on the sinking-fund method is plotted in Fig. 6-2. The rate of depreciation, which equals the rate of growth of principal in the fund, is accelerated as time elapses. This is contrary to the assumed condition, and this curve is the reverse of the book-value curve of Fig. 6-1. Although the sinking-fund method of allocating depreciation may appear to be sound from a financial point of view, its results are unrealistic, and business owners are generally averse to its use. Moreover, although we are assuming for theoretical purposes that the purchase price of an asset will also be the purchase price of its replacement, such a condition is highly improbable. When we take this fact into consideration, the justification for the use of the sinking-fund method loses much of its validity.

Example 6-2. (P.E. examination problem.) A factory is constructed at a first cost of $8,000,000 and with an estimated salvage value of $200,000 at the end of 25 years.

   a. Find its appraisal value to the nearest $100 at the end of 10 years by the sinking-fund method of depreciation, assuming an interest rate of 5 per cent.

   b. Find its appraisal value at the end of 15 years by the straight-line method of depreciation.

Solution.   Part a

Book value = purchase price − principal in sinking fund at end of 10th year

$$S_n = \$8,000,000 - \$200,000 = \$7,800,000$$
$$n = 25 \qquad r = 10 \qquad i = 5 \text{ per cent}$$

By Eq. (4-15),

$$S_{10} = S_{25} \frac{I_{10}}{I_{25}} = 7,800,000 \left(\frac{0.62889}{2.38635}\right)$$
$$= \$2,055,600$$

Book value = $8,000,000 − $2,055,600 = **$5,944,400**

Solution.   Part b

$$\text{Annual depreciation} = \frac{\$7,800,000}{25} = \$312,000$$

Book value at end of 15th year = $8,000,000 − 15 × $312,000 = **$3,320,000**

6-4. Fixed-percentage (Declining-balance) Method. An attempt to make the mathematical procedure for calculating depreciation conform to the decelerating characteristic of depreciation has resulted in the development of the "fixed-percentage" or "declining-balance" method, under which the depreciation occurring each year is considered to be a fixed percentage of the book value at the beginning of the year. For example, if an asset is purchased for $16,000 and is assumed to depreciate by 15 per cent each year, the depreciation charge for the first year is 16,000(0.15), or $2,400; this leaves a book value of $13,600 at the end of the year. The depreciation charge for the second year is 13,600(0.15), or $2,040; the book value at the end of the second year is therefore $11,560. Thus the depreciation charge is maximum during the first year and then gradually tapers off.

In practice, the percentage to be applied in calculating depreciation is not a known quantity; its magnitude must be such as to cause the book value to diminish from the original purchase price of the asset to the salvage value at date of retirement.　Let

C_0 = purchase price of asset
C_r = book value of asset at end of rth year
h = percentage used in calculating depreciation
L = scrap value of asset $(= C_n)$
n = number of years in life span of asset

We then have the following:

Book value at beginning of first year = C_0
Depreciation for first year = $C_0 h$
Book value at end of first year = $C_1 = C_0 - C_0 h$

or $\qquad\qquad C_1 = C_0(1 - h)$
Similarly, $\qquad\quad C_2 = C_1(1 - h) = C_0(1 - h)^2$
　In general,

$$C_r = C_0(1 - h)^r \qquad (6\text{-}3)$$

and $$L = C_n = C_0(1 - h)^n \qquad (6\text{-}3a)$$

Then $$(1 - h)^n = \frac{L}{C_0}$$

or $$1 - h = \left(\frac{L}{C_0}\right)^{1/n} \qquad (6\text{-}4)$$

Expressing this relationship in logarithmic form, we obtain

$$\log (1 - h) = \frac{1}{n} \log \left(\frac{L}{C_0}\right)$$

or $$\log (1 - h) = \frac{1}{n} (\log L - \log C_0) \qquad (6\text{-}5)$$

Example 6-3.　An asset costing \$5,000 has a life expectancy of 6 years and an estimated salvage value of \$800.　Calculate the depreciation charge for each year, using the fixed-percentage method.

Solution

$$C_0 = \$5,000 \qquad L = \$800 \qquad n = 6$$
$$\log 800 = 2.90309$$
$$\log 5,000 = \underline{3.69897}$$
$$\text{Difference} = \overline{9.20412} - 10 = 59.20412 - 60$$
$$\log (1 - h) = 9.86735 - 10$$
$$1 - h = 0.7368$$
$$h = 0.2632 = 26.32 \text{ per cent}$$

Applying this percentage to the successive end-of-year book values, we obtain the results shown in Table 6-2.

TABLE 6-2. DEPRECIATION SCHEDULE FOR FIXED-PERCENTAGE METHOD

Year	Book value at beginning	Depreciation for year	Book value at end
1	$5,000.00	$1,316.00	$3,684.00
2	3,684.00	969.63	2,714.37
3	2,714.37	714.42	1,999.95
4	1,999.95	526.39	1,473.56
5	1,473.56	387.84	1,085.72
6	1,085.72	285.76	799.96

An alternative, and somewhat simpler, method of solution is the following: If C_r and C_{r+1} denote the book values at the end of two consecutive periods, we have

$$C_{r+1} = C_r - C_r h = C_r(1 - h)$$
$$\therefore \log C_{r+1} = \log C_r + \log (1 - h)$$

Hence, the logarithms of the successive end-of-period book values constitute an arithmetical series, having a constant difference of $\log (1 - h)$. The value of this logarithm is given by Eq. (6-5) and is calculated as $9.86735 - 10$ for this particular problem. It is therefore a simple matter to calculate the logarithms of the various book values; the depreciation charge for each year is then obtained by subtracting the book values at the beginning and end of that year. Applying this procedure to the above problem produces the following results:

$$\log C_0 = 3.69897 \qquad C_0 = \$5,000.00$$
$$\log C_1 = 3.56632 \qquad C_1 = 3,684.00$$
$$\log C_2 = 3.43367 \qquad C_2 = 2,714.40$$
$$\log C_3 = 3.30102 \qquad C_3 = 2,000.00$$
$$\log C_4 = 3.16837 \qquad C_4 = 1,473.60$$
$$\log C_5 = 3.03572 \qquad C_5 = 1,085.70$$
$$\log L = 2.90309 \qquad L = 800.00$$

These end-of-year book values are in agreement with the results obtained above. The difference between two successive book values equals the depreciation charge for the intervening year.

Although superficially the fixed-percentage method of calculating depreciation may appear logical, it possesses the following defects:

1. It exaggerates the rate at which depreciation is decelerated as time elapses. According to the results obtained by this method, 31 per cent of the total depreciation for a 6-year period occurs during the first year. This does not appear to be reasonable.

If we refer to Fig. 6-3, which presents the graph of book values for this case, it is seen that the graph has excessive curvature in comparison with the standard graph of Fig. 6-1.

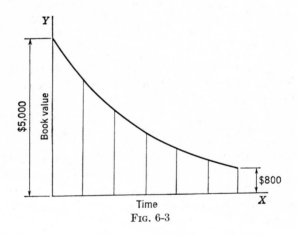

Fig. 6-3

2. This method cannot be applied if no salvage value exists, for a quantity can never be reduced to zero by successive reductions of a fixed percentage. This qualification precludes use of this method in many cases.

3. Moreover, where salvage value does exist, a slight variation of this value will so completely alter the pattern of depreciation charges as to render the results obtained by this method of very dubious value.

In order to illustrate the last point assume that, in the problem solved above, the salvage value is estimated as $1,000 rather than $800. Applying the first method of solution, we find the depreciation factor h to be 23.53 per cent. The annual depreciation obtained in this instance is presented in Table 6-3; for comparison purposes, we have also listed the annual depreciation for the first case, in which the salvage value is $800. The numbers have been rounded off to the nearest ten cents.

TABLE 6-3. COMPARATIVE DEPRECIATION SCHEDULE

Year	Annual depreciation Case 1 (salvage value = $800)	Annual depreciation Case 2 (salvage value = $1,000)	Difference
1	$1,316.00	$1,176.30	$139.70
2	969.60	899.60	70.00
3	714.40	687.90	26.50
4	526.40	526.20	.20
5	387.80	402.20	−14.40
6	285.80	307.80	−22.00
Total..	$4,200.00	$4,000.00	$200.00

Analysis of the results obtained in these two problems discloses certain bizarre features. In the first place, almost 70 per cent of the $200 difference in the total depreciation is reflected in the depreciation charge for the first year of the asset's life, which is certainly a farfetched result. Moreover, a decrease in the total depreciation actually results in an increase in the depreciation charges for the latter years of the asset's life, a condition which is self-contradictory.

These irrational results arise from the following cause: When the salvage value is increased from $800 to $1,000, the depreciation factor h decreases from 26.32 to 23.53 per cent. During the first year, these two percentages are applied to the same book value ($5,000). Hence, at the end of the first year, the book value in Case 1 has declined much more than that in Case 2. When the two depreciation factors are applied to the book values at the commencement of the second year, however, the larger factor is applied against a smaller amount; consequently, the difference between the two depreciation charges for the second year is substantially less than that for the first year. The depreciation charges for the two cases therefore converge until they are approximately equal for one particular year; subsequently, the larger depreciation factor actually produces a smaller charge.

The book-value graphs for the two cases studied above are presented in Fig. 6-4.

Let m be a quantity such that

$$m = \frac{h}{1 - h}$$

Then
$$1 + m = 1 + \frac{h}{1 - h} = \frac{1}{1 - h}$$

$$\therefore \ 1 - h = \frac{1}{1 + m} = (1 + m)^{-1}$$

Hence
$$C_n = C_0(1 - h)^n = C_0(1 + m)^{-n}$$

or
$$C_0 = C_n(1 + m)^n$$

It is thus seen that the book-value curve for the fixed-percentage method is a reversed principal curve; that is, it is reversed in the sense that a given quantity increases by a constant percentage as we retrograde in time. Consequently, from referring to Fig. 6-4, it is evident that the essential feature determining the difference between the annual depreciation charges for the two cases is not the arithmetical difference between the two salvage values but the ratio of one to the other. It is this characteristic that produces the distorted results found above.

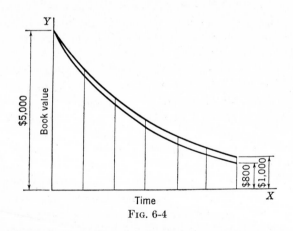

Fig. 6-4

6-5. Double-declining-balance Method. Since its authorization by the Internal Revenue Service in 1954, the so-called "double-declining-balance" method of calculating depreciation has been extensively adopted. Annual depreciation is charged by applying a fixed percentage of the beginning-of-year book value, but the value of this percentage is a function of the life span alone. The formula is as follows:

$$h = \frac{2}{n} \tag{6-6}$$

provided that $n \geq 3$. The designation "double-declining-balance" method arises from the fact that the depreciation charge for the first year as calculated by this method is twice that calculated by the straight-line method.

Since the value of h as given by Eq. (6-6) has no relation to salvage value, the final book value that is obtained by applying h consistently will generally differ from the salvage value, and therefore, a correction is required. If the calculated final book value is less than salvage value, the firm can adhere to the results of the double-declining-balance method up to the point where the book value coincides with salvage value; no depreciation is charged beyond that point. If the calculated final book value exceeds salvage value, the firm has two options. Either it can adhere completely to the results of the double-declining-balance method and then treat the difference between final book value and salvage value as a business loss incurred when the asset is retired, or it can transfer to the straight-line method at some intermediate date. Where the latter course is chosen, the transfer is usually made at the date that will maximize the rate at which depreciation is charged.

Example 6-4. With reference to the asset described in Example 6-3, calculate the depreciation charge for each year if the firm adopts the double-declining-balance method.

Solution

$$C_0 = \$5{,}000 \qquad L = \$800 \qquad n = 6$$

$$h = \frac{2}{6} = 33.33 \text{ per cent}$$

The depreciation charges obtained by applying this rate consistently are calculated in Table 6-4. The calculations are discontinued at the end of the fifth

TABLE 6-4. DEPRECIATION SCHEDULE FOR $h = \frac{1}{3}$

Year	Book value at beginning	Depreciation for year	Book value at end
1	\$5,000.00	\$1,666.67	\$3,333.33
2	3,333.33	1,111.11	2,222.22
3	2,222.22	740.74	1,481.48
4	1,481.48	493.83	987.65
5	987.65	329.22	658.43

year because the computed book value at that point falls below salvage value. Therefore, it is necessary to reduce the depreciation charge for the fifth year. The adjusted depreciation charges are recorded in Table 6-5.

TABLE 6-5. ADJUSTED DEPRECIATION SCHEDULE

Year	Book value at beginning	Depreciation for year	Book value at end
1	$5,000.00	$1,666.67	$3,333.33
2	3,333.33	1,111.11	2,222.22
3	2,222.22	740.74	1,481.48
4	1,481.48	493.83	987.65
5	987.65	187.65	800.00
6	800.00	....	800.00

Example 6-5. An asset costing $10,000 has a life expectancy of 5 years and an estimated salvage value of $300. Calculate the depreciation charge for each year by the double-declining-balance method.

Solution

$$C_0 = \$10,000 \qquad L = \$300 \qquad n = 5$$

$$h = \frac{2}{5} = 40 \text{ per cent}$$

The depreciation charges obtained by applying this rate consistently are calculated in Table 6-6. The final book value of the asset exceeds its salvage value.

TABLE 6-6. DEPRECIATION SCHEDULE

Year	Book value at beginning	Depreciation for year	Book value at end
1	$10,000.00	$4,000.00	$6,000.00
2	6,000.00	2,400.00	3,600.00
3	3,600.00	1,440.00	2,160.00
4	2,160.00	864.00	1,296.00
5	1,296.00	518.40	777.60

Assume that the firm wishes to transfer from the double-declining-balance method to the straight-line method at an intermediate date as a means of bringing the final book value into agreement with salvage value. Selecting each possible transfer date in succession and calculating the uniform depreciation charges for the remaining years of the life of the asset, we obtain the following results:

End of first year: $$\frac{6{,}000 - 300}{4} = 1{,}425.00$$

End of second year: $$\frac{3{,}600 - 300}{3} = 1{,}100.00$$

End of third year: $$\frac{2{,}160 - 300}{2} = 930.00$$

If the transfer is made at the end of the first year, the depreciation charge for the second year will be \$1,425.00 as compared with \$2,400.00 under the double-declining-balance method. Therefore, the end of the first year is not a suitable transfer date. Continuing this analysis, it is seen that the transfer to the straight-line method should be made at the end of the third year. The adjusted depreciation charges are shown in Table 6-7.

TABLE 6-7. ADJUSTED DEPRECIATION SCHEDULE

Year	Book value at beginning	Depreciation for year	Book value at end
1	\$10,000.00	\$4,000.00	\$6,000.00
2	6,000.00	2,400.00	3,600.00
3	3,600.00	1,440.00	2,160.00
4	2,160.00	930.00	1,230.00
5	1,230.00	930.00	300.00

6-6. Modified Fixed-percentage Method. It is possible to modify the fixed-percentage method of allocating depreciation in order to retain its desirable features and at the same time obviate, or at least minimize, its basic defects.

To illustrate the modification, we shall apply the data of Example 6-3, where

$$C_0 = \$5{,}000 \qquad L = \$800 \qquad n = 6$$

In Fig. 6-5, we shall again connect the two points representing the original purchase price and the final salvage value by means of a reversed principal curve, but rather than use line XX as our axis for measuring ordinates to this curve, we shall select an axis located a certain distance below this. In other words, in imagination, we increase each end-of-year book value by a fixed amount in order to calculate the depreciation charge for each year. This will diminish the curvature of the book-value graph. With this modification, the fixed-percentage method becomes applicable even to cases where no salvage value is anticipated. Where salvage value is

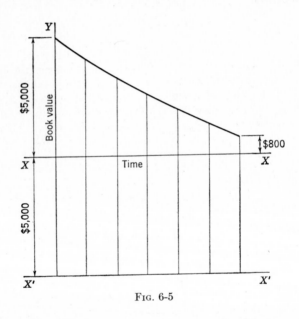

Fig. 6-5

present, a slight modification of this value will not radically alter the pattern of depreciation charges. As to the fixed amount by which the book values should be augmented, no definite answer can be given; but it would appear logical to use the purchase price of the asset for this purpose.

We shall now apply this modified fixed-percentage method to calculate the depreciation charges for the two cases previously discussed. Using the second method of solution presented above, we shall calculate the "modified book value" at the end of each year. This is the sum of the true book value and the original purchase price. The symbol C_r will be employed to denote the modified book value at the end of the rth period.

Case 1

$$\text{Modified purchase price} = \$5,000 + \$5,000 = \$10,000$$
$$\text{Modified salvage value} = \$5,000 + \$\ 800 = \$\ 5,800$$
$$\log 5,800 = 3.76343$$
$$\log 10,000 = 4.00000$$
$$\text{Difference} = \overline{9.76343} - 10 = 59.76343 - 60$$
$$\log (1 - h) = 9.96057 - 10$$

$\log C_0 = 4.00000$	$C_0 = \$10,000.00$
$\log C_1 = 3.96057$	$C_1 = \ \ \ 9,132.00$
$\log C_2 = 3.92114$	$C_2 = \ \ \ 8,339.50$
$\log C_3 = 3.88171$	$C_3 = \ \ \ 7,615.70$
$\log C_4 = 3.84228$	$C_4 = \ \ \ 6,954.70$
$\log C_5 = 3.80285$	$C_5 = \ \ \ 6,351.10$
$\log C_6 = 3.76342$	$C_6 = \ \ \ 5,800.00$

By deducting each book value from the preceding one, we obtain the depreciation charge for the intervening year. These values are shown in Table 6-8. There are also listed, for comparison purposes, the depreciation charges pertaining to Case 2, where the salvage value is $1,000.

TABLE 6-8. COMPARATIVE DEPRECIATION SCHEDULE

Year	Annual depreciation Case 1 (salvage value = $800)	Annual depreciation Case 2 (salvage value = $1,000)	Difference
1	$868.00	$816.00	$52.00
2	792.50	749.70	42.80
3	723.80	688.30	35.50
4	661.00	632.20	28.80
5	603.60	580.50	23.10
6	551.10	533.30	17.80

Evidently, by modifying the fixed-percentage method of calculating depreciation in the manner shown, we have conserved the benefits of its underlying theory while deriving results of a more rational nature.

6-7. Sum-of-integers Method. This method of calculating depreciation, like the fixed-percentage method, is based on the premise that the depreciation process is decelerated with the passage of time. To produce results conforming to this premise, each year's depreciation is calculated by applying a variable fraction to the asset's total wearing value W. The fraction to be used each year is obtained from the following steps:

1. The numerator equals the number of years remaining to the asset's life, plus 1.

2. The denominator equals the total obtained by summing all the integers from 1 up to and including the integer representing the number of years n in the asset's life.

For example, if the asset costs \$6,000 and has a life span of 5 years and no salvage value, then the depreciation charge for the third year is

$$\frac{3}{1+2+3+4+5} \times \$6,000 = \frac{3}{15} \times \$6,000 = \$1,200$$

Let S_n denote the sum of the first n integers; that is,

$$S_n = 1 + 2 + 3 + \cdots + n$$

Reversing the sequence of terms,

$$S_n = n + (n - 1) + (n - 2) + \cdots + 2 + 1$$

Adding the two equations, term for term,

$$2S_n = (n + 1) + (n + 1) + (n + 1) \cdots n \text{ times}$$
or
$$2S_n = n(n + 1)$$
$$S_n = \frac{n(n + 1)}{2}$$

Let D_r denote the depreciation charge for the rth year. Then

$$D_r = \frac{n - r + 1}{n(n + 1)/2} W \tag{6-7}$$

Example 6-6. A machine costing \$10,000 is estimated to have a serviceable life of 8 years, at the end of which time it will have a salvage value of \$1,000. Calculate the annual depreciation, using the sum-of-integers method.

Solution

$$W = \$10,000 - \$1,000 = \$9,000$$
$$n = 8 \quad \therefore \frac{n(n + 1)}{2} = \frac{8 \times 9}{2} = 36$$
$$D_r = \frac{n - r + 1}{n(n + 1)/2} W = \frac{9 - r}{36} \times 9,000 = (9 - r) \times \frac{9,000}{36} = (9 - r) \times 250$$

$$D_1 = 8 \times 250 = \$2,000$$
$$D_2 = 7 \times 250 = 1,750$$
$$D_3 = 6 \times 250 = 1,500$$
$$D_4 = 5 \times 250 = 1,250$$
$$D_5 = 4 \times 250 = 1,000$$
$$D_6 = 3 \times 250 = 750$$
$$D_7 = 2 \times 250 = 500$$
$$D_8 = 1 \times 250 = 250$$
$$\text{Total depreciation} = \$9,000$$

6-8. Depletion. All natural resources of commercial value can be divided into two broad categories. The first group consists of those resources in which nature reproduces, or replaces, the commodity that has been extracted; the second group consists of those resources in which the extracted commodity is not replaced. Thus, in the case of agricultural land, we can assume for valuation purposes that nature will continue indefinitely to yield her annual harvest, barring any event that will render the land infertile. On the other hand, in the case of an oil well, a timber tract, or a mine, the mineral that is extracted is not replaced, and there is consequently a definite limit to the wealth obtainable from the resource. Those in the latter group are designated "wasting" or "depleting" assets, by virtue of the fact that the asset "wastes away" or becomes depleted as it is exploited. In the case of several reproductive assets, the rate of extraction of the commodity must be regulated to coincide with the rate at which nature is capable of replacing it.

6-9. Capital Recovery and Reinvestment of Capital. In order to study the method of evaluating a depleting asset, it is necessary to digress briefly to consider the concept of capital recovery.

Assume that $5,000 is deposited in a fund earning interest at the rate of 6 per cent per annum. The principal and accrued interest remain in this fund until the end of the fourth year, at which time the full amount is withdrawn. At this date, the original principal of $5,000 has expanded to $6,312.40, an increase indicating a total interest earning of $1,312.40. Now, the original principal of $5,000 has earned interest of $300 for 4 years, or a total of $1,200. The remaining $112.40 represents interest earned by the investment of the primary interest of $1,200, that is, through the conversion of the primary interest to principal. In this instance, the depositor receives his interest earnings and retrieves his original capital both at the termination of the investment period.

Now assume that a $5,000 bond, paying interest annually at the rate of 6 per cent, is purchased at par 4 years prior to its maturity. In this case, the investor receives his interest earning of $300 at the end of each year, and at maturity of the bond he retrieves his invested capital. The latter process is known as "capital recovery."

Although both investments considered above earn interest at the rate of 6 per cent, it does not necessarily follow in the second case that the original capital of $5,000 will also amount to $6,312.40 at the end of the fourth year. The final value of the original capital will depend on the rate at which the interest earnings are invested as they are periodically received. This latter rate, however, has no bearing upon the interest rate pertaining to the original investment, which is simply the ratio of interest earned per period to the interest-earning principal.

Assume next that an investor extends a loan of $5,000 for a period of 4 years. At the end of each year, the debtor is to repay one-fourth of the original principal, or $1,250, plus interest on the outstanding debt at the rate of 6 per cent. In this instance, unlike the case of a bond purchased at par, capital recovery is not achieved by a single lump-sum payment at the termination of the investment period; rather, it is achieved gradually through a series of periodic payments extending over the term of the investment. Here again, the invested capital is earning interest at the rate of 6 per cent; however, since the capital is restored periodically to the investor, the amount of invested capital is steadily reduced.

Finally, assume that an investor extends a loan of $5,000, with interest at 6 per cent, the loan to be completely amortized at the end of 4 years by a series of uniform annual payments. The amortization of this loan is recorded in Table 4-5. As we have seen, each payment consists of two elements; the first is a payment of the interest earned by the principal during that period, while the second is a partial payment of principal (capital recovery). By deducting the periodic interest earning from the end-of-year payment, we obtain the following values for capital recovery:

End of year	Capital recovery
1	$1,142.95
2	1,211.53
3	1,284.22
4	1,361.30
Total	$5,000.00

This investment also produces gradual capital recovery during the term of the investment.

Thus, with the mode of achieving capital recovery as our criterion, we can consider that there are basically two distinct types of investment:

Type A. The investor recovers his entire capital at the termination of the investment.

Type B. Through some systematic arrangement of payments, the investor recovers a portion of his capital at the end of each interest period, with the final payment consummating the process of capital recovery.

This classification of investments is concededly too rigid; it recognizes only two extreme types, whereas many investments are of an intermediate nature. For example, when a bond is purchased at a premium, the investor receives at the end of each interest period his interest earning for that period plus a small portion of his invested capital. The bulk of the original capital is restored to him at the redemption of the bond. Nevertheless, despite this defect, it is convenient to maintain this classification for our study of depleting assets.

Where invested capital is recovered periodically, it is improbable that the segments of the original capital that are repaid can be reinvested at the same rate as that earned by the original investment. In the first place, there is a direct relationship between the magnitude of a given sum of money and the investment rate it can command; a sum of money usually cannot be invested at as large a rate as another several times its magnitude. Second, there is an inverse relationship between the magnitude of an investment and the cost per dollar associated with making the investment; because of fixed expenses, the unit cost of investing a small sum of money is greater than the unit cost of investing a large sum. For these reasons, the rate at which capital is periodically reinvested is apt to be less than the rate commanded by the original investment.

Assume that an individual or business firm is confronted with a choice between two investments. Both investments have a term of 10 years. The first requires a capital expenditure of $30,000. Interest at the rate of 7 per cent will be paid annually, and the invested capital will be recovered in its entirety at the termination of the venture. This is a type A investment. The second investment likewise requires a capital of $30,000. At the end of each year, the investor will receive a fixed sum of money; a portion of this sum will represent an 8 per cent return on his investment, while the balance will represent a partial return of capital. The tenth payment will completely restore the invested capital. This is a type B investment.

In view of the fact that the interest rate is 7 per cent for the first investment and 8 per cent for the second, it may appear superficially that the latter investment is more economical. In reality, however, such is not necessarily the case. As we have seen, where capital recovery occurs in periodic stages, there is likely to be a disparity between the investment and reinvestment rates. Assume in the above example of an 8 per cent investment that the segments of capital, as they are recovered, can be reinvested at a rate of only 4 per cent. In this case, the average investment rate earned by the original sum of $30,000 over a 10-year period is substantially less than 8 per cent. It is now evident that the 7 per cent investment, in which the invested capital remains intact until the venture terminates, is the more economical of the two.

Whenever an investment is contemplated, the prospective investment must be evaluated on a relative basis; it must be judged in comparison to alternative available investments. Assuming, as in the illustration above, that two alternative investments require the same initial capital expenditure and are of equal duration, then the proper choice between the two is determined by the answer to the following question: Which investment will produce a greater expansion of the original capital? To rephrase the question, which investment will earn a greater average investment rate? It is thus apparent that in evaluating a type B investment it is necessary to study the complete history of the original capital during the term of the investment. This requires a consideration of the manner in which each segment of capital is reinvested following its recovery. For comparative purposes, it is only the average investment rate that has any true significance.

In undertaking the computation of an average investment rate, we find ourselves confronted with a rather intricate problem. In the first place, at the date the original investment is undertaken, the reinvestment rate is not known; it can only be estimated. Second, it is possible that each segment of capital that is recovered periodically will itself be reinvested in a type B investment, which compounds the difficulty of arriving at an average investment rate. It thus becomes evident that some simplifying assumption must be made regarding the reinvestment of capital in order to develop a rational method of calculating the average investment rate. The nature of this simplifying assumption will vary for each situation.

6-10. Valuation of a Depleting Asset. When the purchase of a depleting asset such as an ore deposit is contemplated, its accurate appraisal requires an estimate of the total quantity of mineral present, the rate at which it can be extracted, and the cost involved. In many numerical problems, the rate of extracting the ore and the cost of this operation are considered to remain constant during the life of the asset, notwithstanding the fact that such is generally not the case. For convenience, we shall assume that the profit realized through exploitation of the asset is calculated annually.

A wasting asset may possess some residual value after it is fully exhausted. For example, when the timber has been completely removed from a tract of land, the land itself has commercial value. In problems pertaining to depletion, if no mention of salvage value appears, it is understood that none exists.

The final exhaustion of a depleting asset produces an abrupt cessation of the income derived from its exploitation, assuming the absence of salvage value. Since none of the invested capital is recovered at the conclusion of the venture, it follows that capital recovery occurred periodically while the asset was being exploited. Hence, the annual profit of a depleting asset consists of two distinct elements; the first is a return on the capital invested in the enterprise during that year, and the second is a partial restoration of the capital originally invested. The purchase and exploitation of a depleting asset therefore form a type B investment and require the determination of an average investment rate rather than a constant rate pertaining to this asset alone.

To provide us with a rational means of calculating this average investment rate, the assumption is made that the owners defer the recovery of their invested capital until the conclusion of the business venture by depositing a fixed portion of the annual profit in a sinking fund, the principal in the fund to equal the original capital investment when the venture terminates. The fund is closed and this principal withdrawn by the owners at that date. Thus, the sum annually deposited in the fund represents capital recovery; the remainder of the annual profit represents a return on the original investment, and is taken by the owners in the form of dividends. In order to achieve a conservative calculation of the average investment rate, it is customary to use a rather low investment rate as the interest rate of the sinking fund.

We are in reality dealing with two distinct investments, namely, the purchase and exploitation of the depleting asset, and the reinvestment of recovered capital in a sinking fund. For simplicity, however, the two investments are considered to constitute a single composite investment, as though it were legally binding upon the owners to defer the recovery of their capital by establishing this sinking fund. By adopting this simplifying assumption, we have transformed the actual type B investment to an equivalent one of type A, in which capital recovery is achieved by means of a single lump-sum payment at the termination of the investment period. Conforming to the terminology generally employed, we shall refer to the purchase of the depleting asset and the reinvestment of capital in the sinking fund as simply the "investment in the asset" and shall refer to the average investment rate pertaining to this composite investment as simply the "investment rate" of the asset.

If a depleting asset does possess salvage value, then this value represents partial capital recovery at the termination of the venture. Therefore, the sinking fund need only accumulate a principal equal to the difference between the invested capital and the salvage value.

We shall now derive an equation for calculating the capital investment to be made in a depleting asset, in correspondence to a given investment rate. Let

C = capital investment
L = salvage value
R = annual net profit (assumed to remain constant)
n = number of years of the business venture
r = rate of return on the investment
i = interest rate earned by the sinking fund

Then $\qquad Cr$ = annual return on the investment
and $\qquad R - Cr$ = annual deposit in the sinking fund

At the end of n years, the principal in the sinking fund must equal the invested capital less the salvage value. Hence, by Eq. (4-1),

$$C - L = (R - Cr) \frac{(1 + i)^n - 1}{i}$$

Solving for C, we obtain

$$C = \frac{R}{r + \dfrac{i}{(1 + i)^n - 1}} + \frac{L}{r \dfrac{(1 + i)^n - 1}{i} + 1} \qquad (6\text{-}8)$$

$$ C = \frac{R}{r + R'_n} + \frac{L}{rs_{\bar{n}} + 1} \qquad (6\text{-}8a) $$

or

Example 6-7. It is estimated that a timber tract will yield an annual profit of $100,000 for 6 years, at the end of which time the timber will be exhausted. The land itself will then have an anticipated value of $40,000. If a prospective purchaser desires a return of 8 per cent on his investment and can deposit money in a sinking fund at 4 per cent, what is the maximum price he should pay for the tract?

Solution

$$ L = \$40,000 \qquad R = \$100,000 \qquad n = 6 $$
$$ r = 8 \text{ per cent} \qquad i = 4 \text{ per cent} $$

By Table B-9, $s_{\bar{6}} = 6.63298$ and $R'_6 = 0.15076$. Substituting in Eq. (6-8a),

$$ C = \frac{100,000}{0.08 + 0.15076} + \frac{40,000}{0.08 \times 6.63298 + 1} $$
$$ = \$459,484 $$

That this purchase price will actually produce an investment rate of 8 per cent can be verified in the following manner:

Capital invested	$= \$459,484$
Salvage value	$= \underline{\ \ 40,000}$
Principal required in sinking fund at end of 6th year	$= \underline{\$419,484}$
Annual deposit required $= 419,484R'_6$	
$= 419,484(0.15076)$	$= \$\ \ 63,241$
$\therefore$ Annual return on investment	
$= \$100,000 - \$63,241$	$= \$\ \ 36,759$

$$ \text{Rate of return} = \frac{36,759}{459,484} = 8 \text{ per cent} $$

Although for mathematical reasons we have blended the purchase of the asset and the deposit of capital in a sinking fund into a single investment, we must not lose sight of the fact that in reality they are two distinct investments and that 8 per cent represents the average rate of return of the two over the 6-year period. (This is sometimes referred to as the speculative rate, to distinguish it from the sinking-fund rate.) The rate of return resulting solely from purchase of the tract is greatly in excess of 8 per cent.

This method of allowing for depletion by applying the device of a hypothetical sinking fund for the investment of recovered capital pending termination of the business venture is closely parallel, in its underlying theory, to the sinking-fund method of allocating depreciation. The depreciation charge that is periodically entered in the accounting records of a business concern serves a dual purpose. First, it records the loss represented by the disintegration of the capital invested in the assets of the enterprise, as these assets deteriorate with use. Second, by deducting this depreciation from the gross profit for that particular period, the firm is in effect reserving a portion of the gross profit for the eventual replacement of its assets, since the enterprise is intended to continue indefinitely. The sinking-fund method thus postulates that the replacement capital so accumulated should not be employed by the firm in the form of working capital, where it is surrounded by the hazards that inhere in all business activity, but rather replacement funds should be maintained intact through some conservative form of investment, such as the creation of a sinking fund. (Implicit in this reasoning is an assumed equality between the original cost of the assets and their replacement cost.) By analogy, it is possible to regard the purchase and exploitation of a timber tract as constituting not a single isolated investment but rather one integral unit of a continuing program of investment in depleting assets. Thus, as one tract is exhausted, it is replaced in this investment program by the purchase of a new tract. If we apply this conception, it is seen that, since the invested capital vanishes as the timber is removed, it is essential that a portion of the gross profit for each period be retained to replace the liquidated capital, thereby permitting the purchase of the succeeding tract. Hence, whether we are dealing with the depreciation of an asset resulting from its use in production or with the depletion of a wasting asset resulting from the extraction of its contents, the creation of a sinking fund can be regarded as a warranty for the preservation of replacement capital.

The preceding analysis, of course, is a theoretical one, designed for the purpose of developing a simple method of evaluating a depleting asset. In actual practice, since entrepreneurs seek constantly to secure the maximum possible return on their available capital, this hypothetical sinking fund is generally not established. Where a depleting asset is owned by a corporation, the law permits dividends to be paid for the full amount of the periodic earnings. Thus, each dividend received by a stockholder comprises both a return on his investment and a partial restoration of his capital, although the stockholder usually lacks knowledge of the precise composition of his dividend.

Many examination problems pertaining to depletion require the calculation of the "purchase price" of an asset that corresponds to a given speculative rate. It is to be understood, however, that this term is being used in a very broad sense, as though it were synonymous with the actual investment. The total investment in a depleting asset represents, basically, the total capitalization of the firm; it encompasses the sums expended both in the purchase of the asset itself and in the purchase of the operating equipment. The periodic profit referred to in these problems denotes the actual profit realized through the sale of the commodity, prior to entering an allowance for depletion. The residual profit after this allowance has been made represents the interest earned by the invested capital. The terminology employed is therefore somewhat contradictory, for in ordinary commercial parlance the term "profit," as applied to a business not dealing in wasting assets, denotes the profit remaining after the allowance for depreciation of the firm's assets has been deducted.

Example 6-8. A mine is purchased for $1,000,000, and it is anticipated that it will be exhausted at the end of 20 years. If the sinking-fund rate is 4 per cent, what must be the annual return from the mine to realize a return of 7 per cent on the investment?

Solution

$$C = \$1,000,000 \qquad L = 0 \qquad n = 20$$
$$r = 7 \text{ per cent} \qquad i = 4 \text{ per cent}$$

By Table B-9, $R'_{20} = 0.03358$. Substituting in Eq. (6-8a),

$$1,000,000 = \frac{R}{0.07 + 0.03358}$$
$$R = 1,000,000(0.10358)$$
$$= \$103,580$$

Example 6-9. (P.E. examination problem.) A syndicate wishes to purchase an oil well which, estimates indicate, will produce a net income of $200,000 per year for 30 years. What should the syndicate pay for the well if, out of this net income, a return of 10 per cent on the investment is desired and a sinking fund is to be established at 3 per cent interest to recover this investment?

Solution

$$R = \$200,000 \qquad L = 0 \qquad n = 30$$
$$r = 10 \text{ per cent} \qquad i = 3 \text{ per cent}$$

By Table B-7, $R'_{30} = 0.02102$. By Eq. (6-8a),

$$C = \frac{200,000}{0.10 + 0.02102} = \frac{200,000}{0.12102}$$
$$= \$1,652,620$$

PROBLEMS

6-1. An asset whose initial cost is $14,000 has an estimated salvage value of $800 and a life expectancy of 12 years. Calculate the depreciation charge for the fourth year and the book value at the end of that year, if depreciation is computed by:

    a. The straight-line method.   *Ans.* $1,100; $9,600

    b. The sinking-fund method, using a 4 per cent rate.   *Ans.* $988; $10,270

    c. The fixed-percentage method.   *Ans.* $1,452; $5,391

    d. The sum-of-integers method.   *Ans.* $1,523; $6,892

6-2. An asset having an initial cost of $10,000 will be retired at the end of 7 years without salvage value. Using the sum-of-integers method, construct a depreciation schedule, recording the annual depreciation and end-of-year book value.

6-3. An asset costing $60,000 had a life expectancy of 10 years and an estimated salvage value of $5,000. Depreciation was computed by the fixed-percentage method. At the beginning of the seventh year, however, the estimated salvage value was revised to $2,000, and the percentage applied to calculate depreciation for the remaining life of the asset was increased accordingly. Calculate the book value of the asset at the end of the eighth year. Construct a depreciation schedule by calculating the two percentages used in the calculations.

6-4. (P.E. examination problem.) It is estimated that the Deep Gulch Mine, which is currently operating, can be expected to make a net profit, after all taxes are paid, of $150,000 per year for 35 years, at which time it will be exhausted and have no salvage value. What can you afford to pay for the mine now, so that you will have an annual income of 12 per cent on your investment after you have made an annual deposit into a fund which, at 3 per cent interest, will accumulate to the amount of your investment (return of investment) in 35 years, when the mine will be exhausted?

6-5. It is estimated that a mine will yield an annual profit of $30,000 for 10 years, at the end of which time it will have a salvage value of $4,000. Assuming that recovered capital can be reinvested at 3 per cent, what is the purchase price required to yield an investment rate of 7 per cent?   *Ans.* $193,013

6-6. Verify the solution for the preceding problem by computing the annual return on the investment.

6-7. A depleting asset was purchased for $800,000, and had a residual value of $50,000 when it became exhausted at the expiration of 15 years. The sinking-fund rate was 4½ per cent. If the venture yielded an 8 per cent return, what was the annual profit?

Solution

$$
\begin{array}{lcr}
\text{Annual return on investment} & = & \$\ 64,000 \\
\text{Annual deposit in fund} & = & 36,083 \\
\text{Annual profit} & = & \$100,083 \\
\end{array}
$$

6-8. A mine was purchased for $1,300,000, and estimates indicate that it will be exhausted at the expiration of 25 years without salvage value. What annual profit is required if the entrepreneurs are to realize a 10 per cent return on their investment, with a sinking-fund rate of 3½ per cent?

6-9. The purchase price of a depleting asset is $500,000, and it is expected to yield a net profit of $75,000 for 10 years. If the sinking-fund rate is 3 per cent, what salvage value must the asset possess for the investment to yield a return of 6½ per cent?   *Ans.* $12,785

6-10. A timber tract was purchased for $2,000,000, and yielded an annual profit of $300,000 for 12 years. Upon cessation of operations, the land was sold for $400,000. Based on a sinking-fund rate of 3½ per cent, what was the investment rate pertaining to this venture?   *Ans.* 9.52 per cent

6-11. An oil well was expected to produce an annual profit of $250,000 for 20 years, at the end of which time it would be valueless. On the basis of this estimate, it was purchased at a price calculated to yield an 8 per cent return on the investment, with a sinking fund rate of 3 ½ per cent. However, the operating costs proved to be greater than anticipated, reducing the annual profit to only $210,000. What was the true investment rate?   *Ans.* 6.154 per cent

It is interesting to note that while the annual profit declined by 16 per cent, the annual return on the investment declined by 23 per cent. This disparity arises from the fact that annual capital recovery, which consumes a large portion of the profit, remains constant in amount.

6-12. In the preceding problem, what is the true investment rate if the annual profit is $280,000? On a percentage basis, how does the increase of annual return compare with the increase of annual profit?

CHAPTER 7

COST COMPARISON OF ALTERNATIVE METHODS

7-1. Introduction. When a business firm or governmental agency must perform some necessary task, whether it be manufacturing a standard commodity, repairing equipment, or building a new water-treatment plant, there are generally several alternative methods by which the task can be performed. For example, there are alternative types of manufacturing processes, varying degrees of repair, alternative building materials, etc. Thus, associated with every necessary task is a set of alternative methods of performing that task.

In order to select the method of optimal efficiency, it is necessary among other things to compare the costs of the alternative methods. We shall consider three methods of making this cost comparison: the annual-cost method, the present-worth method, and the capitalized-cost method.

7-2. Equivalent Annuity. Consider an annuity consisting of 10 annual payments of **$1,000** each. If money is worth 5 per cent, then the values of this annuity at its origin and terminal dates, respectively, are

$$_{10}T_0 = 1,000(7.72173) = \$7,721.73$$
$$S_{10} = 1,000(12.57789) = \$12,577.89$$

To obtain the value of the annuity at some intermediate date, such as the end of the sixth year, we have

Value of 6 past payments = $1,000(6.80191)$	=	$ 6,801.91
Value of 4 future payments = $1,000(3.54595)$	=	3,545.95
Total value of annuity	=	$10,347.86

The three values of the annuity calculated above, of course, are linked by the basic money-time relationship of Eq. (3-2).

Thus, with a stipulated interest rate, it is possible to convert a given annuity to an equivalent single payment made at any date whatsoever. If we reverse the process, it is possible to convert a single payment to an equivalent annuity having specified origin and terminal dates. For example, assume we wish to convert a payment of $1,000 made on Jan. 1, 1950, to an equivalent annuity consisting of five end-of-year payments, with the origin date of the annuity being Jan. 1, 1950 and the conversion being based on an interest rate of 6 per cent. The payment of $1,000 represents the value of the annuity at the beginning of 1950, which is also the origin date of the annuity. Hence, the annual payment is

$$R = 1,000 \left[\frac{0.06}{1 - (1.06)^{-5}} \right] = 1,000(0.23740)$$
$$= \$237.40$$

Assume now that the single payment of $1,000, made at the beginning of 1950, is to be converted to an equivalent annuity of five annual payments, whose origin date is Jan. 1, 1948. Since $1,000 represents the combined value of the two past payments and the three future payments, we have

$$1,000 = R \frac{(1.06)^2 - 1}{0.06} + R \frac{1 - (1.06)^{-3}}{0.06}$$
$$= R(2.06000) + R(2.67301) = R(4.73301)$$
$$R = \frac{1,000}{4.73301} = \$211.28$$

An alternative method of solution is to convert the payment of $1,000 to its equivalent value at either the origin or terminal date of the annuity and then to calculate the annual payment accordingly. Using the terminal date of the annuity for this purpose, we obtain

$$S_5 = 1,000(1.06)^3 = \$1,191.02$$
$$R = 1,191.02 \left[\frac{0.06}{(1.06)^5 - 1} \right] = 1,191.02(0.17740)$$
$$= \$211.28$$

7-3. Annual Cost. When an asset is used in the operation of a business, it is often desirable for comparative purposes to replace the actual expenditures associated with the asset with a hypothetical equivalent annuity consisting of end-of-year payments. The origin date of this annuity is the purchase date of the asset, and its terminal date is the date of retirement of the asset. The periodic payment of this equivalent annuity is referred to as the "equivalent uniform annual cost."

Example 7-1. A machine was purchased for $12,000 at the beginning of 1960. Its operating cost amounted to $500 for each year it was in operation. (For convenience, the operating cost of $500 can be regarded simply as a lump-sum payment made at the end of each year.) Major repairs, amounting to $4,000, were required at the end of 1970. The machine was scrapped without salvage value at the end of 1976. What was the equivalent uniform annual cost of this machine? Use an interest rate of 5 per cent.

Solution. The annual operating costs already constitute an annuity and therefore require no conversion. To convert the intermediate payment of $4,000, we shall translate this amount to its equivalent value at the date of purchase. We then obtain

$$\text{Equivalent uniform annual cost} = 12,000 \left[\frac{0.05}{1 - (1.05)^{-17}} \right]$$

$$+ 4,000(1.05)^{-11} \left[\frac{0.05}{1 - (1.05)^{-17}} \right] + 500$$

$$= 12,000 R_{17}'' + 2,338.80 R_{17}'' + 500$$

$$= 14,338.80(0.08870) + 500$$

$$= 1,271.85 + 500$$

$$= \$1,771.85$$

As seen in the solution of this problem, the intermediate payment of $4,000, after being translated to its equivalent value at the purchase date of the asset, can be added directly to the purchase price. The single sum thus obtained is then converted to an equivalent annuity.

Example 7-2. A machine was purchased for $15,000 and was retired at the end of 10 years with a salvage value of $2,000. The annual operating cost was $700. Based on an interest rate of 6 per cent, what was the equivalent uniform annual cost?

Solution. Since the salvage value represents income, the sum of $2,000 is to be converted to an equivalent series of uniform end-of-year receipts, extending over the life of the machine. The salvage value represents the value of these receipts at their terminal date. The equivalent annual receipt is then deducted from the elements of cost.

$$\text{Equivalent uniform annual cost} = 15,000 \left[\frac{0.06}{1 - (1.06)^{-10}} \right]$$

$$+ 700 - 2,000 \left[\frac{0.06}{(1.06)^{10} - 1} \right]$$

$$= 15,000 R_{10}'' + 700 - 2,000 R_{10}'$$

$$= 15,000(0.13587) + 700 - 2,000(0.07587)$$

$$= 2,038.05 + 700 - 151.74$$

$$= \$2,586.31$$

As an alternative method of calculating this cost, we can translate the salvage value to its equivalent value at the purchase date and deduct this amount from the purchase price. We shall refer to the difference between the two as the "effective purchase price," which in this case is being evaluated at the purchase date. This single sum is then converted to an equivalent annuity. Applying this method, we obtain

Value of effective purchase price at date of purchase

$$= 15,000 - 2,000(1.06)^{-10}$$
$$= 15,000 - 1,116.78$$
$$= \$13,883.22$$

$$\text{Equivalent uniform annual cost} = 13,883.22(0.13587) + 700$$
$$= 1,886.31 + 700$$
$$= \$2,586.31$$

A modification of this method is to translate the purchase price to its equivalent value at the retirement date of the machine, from which the salvage value is then deducted. The difference represents the value of the effective purchase price at the retirement date and also represents the value of the equivalent annuity at its terminal date. This procedure yields the following:

Value of effective purchase price at date of retirement

$$= 15,000(1.06)^{10} - 2,000$$
$$= 26,862.75 - 2,000$$
$$= \$24,862.75$$

$$\text{Equivalent uniform annual cost} = 24,862.75(0.07587) + 700$$
$$= \$2,586.34$$

The expression "equivalent uniform annual cost" will henceforth be contracted to the simpler form "annual cost."

7-4. Selection of Interest Rate for Calculating Annual Cost. The annual cost pertaining to a particular asset is of course a function of the interest rate used in the calculations. Hence, it is of prime importance that we establish a definite criterion for determining the rate on which the calculations are to be based. In each instance, the selection of an appropriate interest rate requires a consideration of the purpose underlying the computation of the annual cost of the asset.

Consider the following simple problem: A machine was purchased for $10,000 and retired at the end of 10 years without salvage value. The annual operating cost was $800.   By installing this machine, the business firm reduced its annual labor cost by $2,000.   (For simplicity, this can be regarded as a single sum having this value at the end of the year.) Compute the annual cost of this machine to determine whether its use was justified from a monetary viewpoint.

Since the sums expended for this machine total $18,000, while the labor savings resulting from its installation total $20,000, it follows that the investment of capital in this asset did produce a profit for the business firm.   The question to be answered, however, is whether this profit was sufficient, for there are always alternative investments available.   For example, assume that at the date of purchase of this machine the firm had a choice between expending $10,000 for its purchase or undertaking a different investment of $10,000 for a term of 10 years.   The latter investment would have provided uniform annual dividends, each dividend consisting of a 7 per cent return on invested capital plus a partial recovery of capital.   To determine the amount of the annual dividend, we have

$$R = 10{,}000R''_{10} = 10{,}000(0.14238)$$
$$= \$1{,}423.80$$

Therefore, by investing its capital in this machine in preference to undertaking the alternate investment available, the firm has both deprived itself of an annual income of $1,423.80 and incurred an annual expense of $800 for its operation.   Totaling these amounts, we obtain $2,223.80 as the annual "cost" of the asset to the firm.

If money did not possess a time value (or, to express this in another manner, if the investment rate were zero rather than 7 per cent), the annual dividend would simply be $1,000 rather than $1,423.80, producing a total annual cost of $1,800.   It is thus apparent that by taking cognizance of the time value of money in our calculation of annual cost, we are including in that cost not only the sums actually expended in the purchase and operation of the asset but also the potential interest income of $423.80 annually that has been forfeited as a consequence of the expenditure of capital for its purchase.   When we convert the actual disbursements to an equivalent annuity on the basis of a 7 per cent interest rate, we are implicitly recognizing that each disbursement represents a loss both of the sum expended and of its latent future interest, calculated at the rate of 7 per cent.   Evidently, the appropriate interest rate to be used in the calculation of annual cost in this particular case is determined by the answer to the latter part of this question: What investment was available to the business enterprise as an alternative to purchase of this asset, and what was the investment rate applying thereto?

Since the annual cost of the machine was $2,223.80 and the annual saving resulting from its use only $2,000, then purchase of this machine was not a wise investment. The fact that annual cost exceeded the annual saving does not imply that there was an actual monetary loss incurred by purchasing the machine, for such was not the case. The excess of annual cost over annual saving simply discloses that the profit realized by purchase of this machine was insufficient; the rate of return on this investment fell below the 7 per cent rate obtainable elsewhere. When an investment is evaluated by determining its annual cost, it is in effect being evaluated not on an individual basis but in comparison with other available investments. On a relative basis, this investment has not proved justified.

Now assume that the annual reduction of labor cost resulting from the use of this machine was $2,500. This indicates that the rate of return on this investment exceeded 7 per cent, and the investment was therefore warranted. In each instance, unless we make additional calculations, we lack knowledge of the precise rate of return from the application of capital toward purchase and operation of this machine. The comparison of annual cost to annual saving simply discloses whether this rate falls above or below the 7 per cent rate which we are employing as a standard.

Consider next the following problem: A business firm is contemplating the installation of a machine to perform a certain function, and two different types are under consideration. Both machines have an estimated life of 15 years, with no prospect of salvage value. Machine A has a purchase price of $20,000 and an annual operating cost of $1,300. Machine B has a purchase price of $16,000 and an annual operating cost of $1,750. Calculate the annual cost of each machine to determine which is more economical.

Assume that, if neither of these two machines is purchased, the available capital can be invested to yield a return of 7 per cent. In this problem, this fact does not enter directly into the problem, for the choice is restricted to the purchase of machine A or machine B. Any other form of investment is precluded by the conditions of the problem. There is, however, a difference of $4,000 between the purchase price of the two machines. If machine B is purchased in preference to machine A, this

$4,000 is available for some other form of investment, and the income obtainable therefrom must be taken into consideration in a comparison of the two machines. Since there is some relation between the quantity of invested capital and the interest rate it can command, we shall assume that the investment of $4,000 can produce a return of only 6 per cent. This, then, is the rate to be used in calculating the annual cost of each asset. For the present case, therefore, the selection of the appropriate interest rate is determined by the answer to this question: If the asset having the lower first cost is purchased, at what rate can the difference between the purchase price of each asset be invested?

Machine A:

$$\text{Annual cost} = 20{,}000 \left[\frac{0.06}{1 - (1.06)^{-15}} \right] + 1{,}300$$
$$= 20{,}000 R''_{15} + 1{,}300$$
$$= 20{,}000(0.10296) + 1{,}300$$
$$= 2{,}059.20 + 1{,}300$$
$$= \$3{,}359.20$$

Machine B:

$$\text{Annual cost} = 16{,}000(0.10296) + 1{,}750$$
$$= 1{,}647.36 + 1{,}750$$
$$= \$3{,}397.36$$

Machine A is the more economical of the two.

It may be objected that, although the 6 per cent rate applies only to the difference of $4,000 between the purchase price of the two assets, we are applying this rate to the entire purchase price. However, since we are only interested in comparing the annual cost of one asset with the other, the objection loses its validity. It is only the numerical difference between the two annual costs that concerns us in our present study, not the absolute value of each.

This problem of determining the economic choice between two different assets can be made analogous to the problem preceding it if it is analyzed in the following manner: By purchasing machine A in preference to machine B, the firm is in effect investing $4,000 at the purchase date in order to secure 15 annual returns of $450 each (the difference between the operating cost of each asset). This $4,000 capital can also be invested elsewhere in a manner that yields 15 uniform annual dividends, each dividend consisting of a 6 per cent return on the investment plus a partial recovery of capital. The amount of the annual dividend is

$$4,000(0.10296) = \$411.84$$

Thus, by purchasing machine A in preference to machine B, the firm is depriving itself of a potential annual income of \$411.84 in order to achieve an annual saving of \$450 in operating cost. The purchase of machine A is therefore feasible.

The fundamental fact regarding annual cost is that it is a relative rather than absolute cost.

7-5. Constancy of Annual Cost. In the problems solved above, the annual cost which we have calculated is the cost pertaining to one particular life of an asset. However, our study of finance is based on the assumption that the economic and technological conditions existing at the instant of time under consideration remain permanent. From this premise it follows that, if the asset is to be replaced by another of the same type, then the life and cost pattern of this second asset, and all subsequent assets, will be identical with the original one. For convenience, we can regard each replacement of an asset as representing a new life of the same asset. It also follows as a consequence of the assumption of static conditions that the interest rate used in the calculation of annual cost will remain permanent. Therefore, although we have calculated annual cost for only one particular life of an asset, we assume that this annual cost will remain constant and be applicable to all future lives as well. In a particular problem, if we are to compare the annual costs pertaining to two assets whose life spans are unequal, this difference does not invalidate the basis of comparison, for the annual cost of each asset is considered to be permanent.

As emphasized previously, by assuming static economic conditions we are not deluding ourselves regarding the true dynamic nature of our economy; we are simply making such an assumption to facilitate a mathematical analysis of financial problems. If the manner in which future economic and technological changes may affect annual cost can be anticipated with a reasonable degree of certainty, then these factors should be taken into consideration in arriving at a business decision.

Example 7-3. (P.E. examination problem.) The butt of a pole in an electric distribution system has decayed to the point where it is necessary either to replace the pole or to "stub" it. A new pole will cost $40 installed and will have an estimated life of 22 years. The remaining life in the upper part of the present pole is estimated to be 9 years, which may be realized if the pole is stubbed. A cedar stub will cost $12. The upper part of the present pole is valued at $6 when used with a stub. The salvage value of a new pole, after 22 years, is placed at $6. With money at 7 per cent, is it cheaper to stub or to replace?

Solution. If we assume that the $6 value of the upper part of the existing pole can be realized if the pole is scrapped, this value represents a portion of the first cost of continuing the present pole in use by stubbing it, since potential income is thus forfeited. We therefore have the following cost data:

	To stub pole	To replace pole
Initial cost............	$18	$40
Life in years...........	9	22
Salvage value..........	...	$ 6

Performing the necessary arithmetical calculations, we arrive at these results:

$$\text{Annual cost if pole is stubbed} = \$2.76$$
$$\text{Annual cost if pole is replaced} = \$3.49$$

The annual cost of $3.49 pertaining to a new pole is assumed to remain constant and hence applicable to an infinite period of time. On the other hand, the annual cost of $2.76 that results if the present pole is stubbed applies only to the 9-year period for which the present life has been extended. However, we are assuming that at the end of 9 years the pole can still be replaced with a new one whose annual cost is $3.49. Therefore, if the present pole is stubbed rather than replaced, there is an annual saving of 73 cents for the next 9 years; after that time, the annual cost will be independent of the course of action pursued at the present date. It is therefore cheaper to stub the pole.

7-6. Annual-cost Equations. We shall now derive equations for annual cost applicable to those cases where all intermediate disbursements are of a uniform nature. Let

P = purchase price, or initial cost, of asset
L = salvage value
n = life of asset in years
i = interest rate
O.C. = annual operating costs

The value of the effective purchase price at the date of purchase is

$$P - L(1 + i)^{-n}$$

Converting this to an equivalent annuity, we obtain

$$\text{Annual cost} = [P - L(1 + i)^{-n}]\frac{i}{1 - (1 + i)^{-n}} + \text{O.C.}$$

$$= \frac{Pi - Li(1 + i)^{-n} + (Li - Li)}{1 - (1 + i)^{-n}} + \text{O.C.}$$

$$= (P - L)\frac{i}{1 - (1 + i)^{-n}} + Li + \text{O.C.} \qquad (7\text{-}1)$$

or $$\text{Annual cost} = (P - L)R_n'' + Li + \text{O.C.} \qquad (7\text{-}1a)$$

An alternative equation for annual cost can be derived by evaluating the effective purchase price at the retirement date of the asset. This method yields the following:

Value of effective purchase price at retirement date $= P(1 + i)^n - L$

Converting this to an equivalent annuity, we obtain

$$\text{Annual cost} = [P(1 + i)^n - L]\frac{i}{(1 + i)^n - 1} + \text{O.C.}$$

$$= \frac{Pi(1 + i)^n - Li + (Pi - Pi)}{(1 + i)^n - 1} + \text{O.C.}$$

$$= \frac{(P - L)i + Pi[(1 + i)^n - 1]}{(1 + i)^n - 1} + \text{O.C.}$$

$$= (P - L)\frac{i}{(1 + i)^n - 1} + Pi + \text{O.C.} \qquad (7\text{-}2)$$

or $$\text{Annual cost} = (P - L)R_n' + Pi + \text{O.C.} \qquad (7\text{-}2a)$$

There are some assets whose lives are considered infinite in duration. By setting n equal to infinity in either of the above equations, we obtain

$$\text{Annual cost for perpetual life} = Pi + \text{O.C.} \qquad (7\text{-}3)$$

Equations (7-1) and (7-2) can be obtained by simple logic, in this manner: assume momentarily that the income from salvage is obtained at the date of purchase rather than the date of retirement. The net payment for the asset would then be $P - L$. This payment can be converted to an equivalent annuity for which the annual payment is

$(P - L)R_n''$. By falsely assuming that the owners obtain an income L at date of purchase, we are crediting them with an imaginary annual interest earning of Li for each year the asset is in existence. Therefore, as a correction, we must add the amount Li to the annual cost of the asset, and Eq. (7-1a) results.

Similarly, assume momentarily that the payment P to purchase and install the asset is made at the date of retirement rather than the date of purchase. The net payment at the date of retirement would be $P - L$. This payment can be converted to an equivalent annuity for which the annual payment is $(P - L)R_n'$. By falsely assuming that the owners retain the capital P during the life of the asset, we are crediting them with an imaginary annual interest earning of Pi. Therefore, as a correction, we must add the amount Pi to the annual cost of the asset, and Eq. (7-2a) results.

Equations (7-1) and (7-2) are sometimes derived by assuming that purchase of the asset is financed with borrowed money. Although the reasoning underlying this method is fallacious, it is nevertheless necessary that readers familiarize themselves with it in order to understand the method when it is encountered.

First, consider that the funds required to purchase the asset are obtained by means of two separate loans, one for the sum $(P - L)$ and another for the sum L. The interest rate for both loans is i. The first loan is to be completely amortized at the retirement date of the asset by a series of uniform annual payments. The second loan stipulates that interest is to be paid at the end of each year, with the principal to be repaid as a lump sum at the retirement date. Since the salvage value L of the asset will provide the funds necessary to repay the principal, the only expense incurred by the firm with regard to this loan is the annual payment of interest.

The end-of-year disbursements attributable to this asset are the following:

1. A payment of $(P - L)R_n''$ to amortize the first loan
2. An interest payment of Li on the second loan
3. The annual operating cost O.C.

The borrowed-money method of calculating annual cost regards this cost as the sum of the actual disbursements made each year, which is a misconception of the true nature of annual cost. Summing the disbursements listed above and equating the total to annual cost, we arrive at Eq. (7-1a).

Next, assume that the sum of money P required to purchase the asset is obtained by a single loan, the terms of the loan requiring the payment of interest at the rate i at the end of each year and repayment of the principal as a lump sum at the retirement date. Assume also that in order to accumulate the principal of the loan by its date of maturity the business firm establishes a sinking fund earning interest at the identical rate i. Since a portion of the funds required to repay the principal will be supplied by the salvage value, the principal to be accumulated in the sinking fund is only $(P - L)$.

The end-of-year disbursements attributable to this asset are the following:

1. An interest payment of Pi
2. A sinking-fund deposit of $(P - L)R'_n$
3. The annual operating cost O.C.

Summing these items, we arrive at Eq. (7-2a).

The borrowed-money method of computing annual cost can be extended to embrace those cases in which intermediate expenses of a nonuniform character occur. For example, assume that at the end of a particular year in the life of an asset a disbursement is required for major repairs. In order to equalize the annual expenditures, the firm can establish a sinking fund whose principal will be sufficient to provide not only for repayment of the principal of the loan at its maturity but also for the withdrawal of funds for this extraordinary expenditure at the intermediate date.

The borrowed-money method, however, possesses the following defects:

1. Its entire approach to the problem is unrealistic. Even if the purchase of the asset is actually financed with borrowed funds, it is scarcely conceivable that the firm can obtain a loan whose term equals in length the life span of the asset. Moreover, the assumption of perpetual life for an asset precludes application of the financing methods presented above.

2. It is highly improbable that the sinking fund will earn an interest rate equal to that which the firm must pay for its borrowed money.

3. The most serious shortcoming of this method is its treatment of annual cost as an absolute rather than comparative cost, which consists exclusively of sums actually expended with regard to the asset. Potential

income that is forfeited through purchase of the asset is not taken into account. This fallacy compels the use of the interest rate pertaining to borrowed money in the calculation of annual cost, although this is generally not the appropriate rate to be used in a particular case.

To illustrate the last point, consider the following problem: A machine was purchased for $10,000 and retired at the end of 8 years without salvage value. The annual operating cost was $1,000. The funds to purchase this machine were obtained by borrowing $10,000, the loan to be completely amortized at the end of 8 years through a series of uniform annual payments with interest at 5 per cent. If the annual reduction of labor cost resulting from installation of this machine was $2,600, was the investment of capital in this machine justified?

The annual expenditures attributable to this machine were the following:

1. A payment to amortize the loan, which at a 5 per cent interest rate amounted to

$$R = 10,000R_8'' = 10,000(0.15472) = \$1,547.20$$

2. The annual operating cost of $1,000

Hence, the total annual disbursements amounted to $2,547.20. Since the annual labor saving was $2,600, a profit was realized by the use of this machine. It does not necessarily follow, however, that the investment was justified, for the borrowed money could also have been invested in some alternative manner. Assume that one such possible investment would have provided a series of eight uniform annual dividends, each dividend representing a return of 8 per cent on invested capital plus a partial recovery of capital. The amount of the annual dividend would then be

$$R = 10,000 \left[\frac{0.08}{1 - (1.08)^{-8}} \right] = 10,000(0.17401)$$
$$= \$1,740.10$$

Thus, by purchasing this machine in preference to undertaking the latter investment, the firm has deprived itself of an annual income of $1,740.10 and has incurred an annual expenditure of $1,000 for operating cost. The annual cost of the asset is therefore $2,740.10, which exceeds the annual saving of $2,600. Purchase of the machine was not justified, since the alternative investment available would have produced greater returns. It is thus clear that the periodic payment required to discharge the debt does not enter into the calculation of annual cost; this amount would be the same irrespective of the particular investment undertaken. In all cases, we return to the basic question concerning the alternative available form of investment and the interest rate applicable thereto.

7-7. Comparisons Based on Annual Cost.   The annual cost of an asset serves as a highly useful tool in arriving at a business decision where an economic choice is to be made among several alternative courses of action. The following examples will illustrate the applicability of annual cost for comparison purposes.   All numerical solutions will be rounded off to the nearest dollar.

Example 7-4.   (P.E. examination problem.)   Compare the following types of construction on the basis of annual cost, using interest at 5 per cent.

	Timber trestle	Steel bridge
First cost................	\$4,000	\$12,000
Life in years.............	15	60
Salvage value............	None	\$1,000
Annual maintenance.......	\$500	\$300

Solution

$$\text{Annual cost} = (P - L)R_n'' + Li + \text{O.C.}$$
$$\text{Annual cost of timber trestle} = 4{,}000R_{15}'' + 500$$
$$= 4{,}000(0.09634) + 500 = 385 + 500$$
$$= \$885$$
$$\text{Annual cost of steel bridge} = 11{,}000R_{60}'' + 1{,}000(0.05) + 300$$
$$= 11{,}000(0.05283) + 1{,}000(0.05) + 300$$
$$= 581 + 50 + 300$$
$$= \$931$$

The timber trestle is therefore the economic choice.

Example 7-5.   (P.E. examination problem.)   The timber floor of a bridge requires replacement.   Experience has shown that an untreated timber floor costs \$3,000 and has a life of 10 years and an annual maintenance charge of \$500.   How much extra will it pay to spend on treatment of the timber if the life of the floor will thereby be increased to 15 years and the annual maintenance charges reduced to \$300?

Solution

	Untreated timber	Treated timber
First cost................	$3,000	P
Life in years............	10	15
Operating cost..........	$500	$300
Salvage value...........	......	....

The interest rate to be used is not specified in this problem; we shall select 7 per cent for our calculations.

Annual cost of untreated timber floor $= 3{,}000R''_{10} + 500 = 427 + 500 = \$927$
Annual cost of treated timber floor $= PR''_{15} + 300$

To determine the maximum allowable value of P, we equate the two annual costs. Then

$$PR''_{15} + 300 = 927$$

or

$$P\,\frac{0.07}{1 - (1.07)^{-15}} + 300 = 927$$

$$P = 627\left[\frac{1 - (1.07)^{-15}}{0.07}\right] = 627\ {}_{15}t_0$$

$$= 627(9.10791)$$

$$= \$5{,}712$$

Allowable incremental investment $= \$5{,}712 - \$3{,}000 = \$2{,}712$

Example 7-6. (P.E. examination problem.) A certain floor surfacing in a factory has to be replaced every 5 years at a cost of $1,500. How long should a floor surfacing costing twice as much last to justify the larger expenditure if a 6 per cent return is required on the investment?

Solution

	Type A	Type B
First cost..............	$1,500	$3,000
Life in years..........	5	n
Salvage value.........	......	......

To determine the minimum value of n, we equate the two annual costs.

$$\text{Annual cost of type A} = \text{annual cost of type B}$$

$$1{,}500R''_5 = 3{,}000R''_n$$

$$R''_n = 0.5R''_5 = 0.5(0.23740)$$

$$= 0.11870$$

$$n = 12.9 \text{ years}$$

Example 7-7. Compare the following machines on the basis of annual cost, using an interest rate of 8 per cent.

	Machine A	Machine B
First cost................	$6,800	$12,000
Salvage value............	......	$1,000
Life in years............	6	10

For machine A, the estimated annual operating cost is $1,300. For machine B, it is $800 for the first 4 years, $1,200 for the next 3 years, and $1,500 for the remaining 3 years.

Solution. We shall apply Eq. (7-1a).

Machine A:

$$\text{Annual cost} = 6,800 R_6'' + 1,300$$
$$= 6,800(0.21632) + 1,300$$
$$= \$2,771$$

Machine B: The annual operating costs are shown in Fig. 7-1a. The payments are nonuniform, but they can readily be converted to an equivalent set of annuities. First, select the end of the tenth year as the valuation date, and resolve the payments for the last 6 years into the sums shown in Fig. 7-1b.

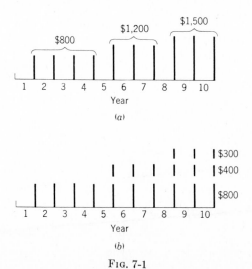

Fig. 7-1

At date of retirement,

$$\text{Value of payments} = 800s_{\overline{10}} + 400s_{\overline{6}} + 300s_{\overline{3}}$$
$$= 800(14.48656) + 400(7.33593) + 300(3.24640)$$
$$= \$15,498$$

$$\text{Equivalent uniform O.C.} = 15,498R'_{10} = 15,498(0.06903)$$
$$= \$1,070$$

Alternatively, select the date of purchase as the valuation date. Assume that the annual operating cost is \$1,500 during the entire life. As a correction, apply an annual saving of \$300 for the first 7 years and, in addition, an annual saving of \$400 for the first 4 years.

At date of purchase,

$$\text{Value of payments} = 1,500 \, _{10}t_0 - 300 \, _7t_0 - 400 \, _4t_0$$
$$= 1,500(6.71008) - 300(5.20637) - 400(3.31213)$$
$$= \$7,178$$

$$\text{Equivalent uniform O.C.} = 7,178R''_{10} = 7,178(0.14903)$$
$$= \$1,070$$

$$\text{Annual cost} = 11,000R''_{10} + 1,000(0.08) + 1,070$$
$$= \$2,789$$

Machine A is slightly more economical.

With reference to machine B, it is instructive to compare the equivalent uniform operating cost with the arithmetical mean operating cost computed without reference to time. The equivalent uniform cost is \$1,070 and the mean cost is \$1,130. Each payment represents a sacrifice both of capital and of the interest that this capital would have earned had it remained in the owners' possession. Therefore, the payments made in the early years of an asset have a more pronounced effect on annual cost than do the payments made in the later years. Since the annual payments in the early years are less than those in the later years, it follows that the equivalent uniform operating cost is less than the mean operating cost.

Example 7-8. A manufacturing firm installed a facility having the following cost data:

First cost.................	\$80,000
Salvage value..............	\$4,000
Life in years..............	9
Annual maintenance........	\$6,800

After the facility had been in operation for 5 years, it was proposed that the facility be improved immediately to extend its life by 2 years, reduce annual maintenance to $5,000 for the remaining life, and increase the salvage value to $7,500. If money is worth 10 per cent, what is the maximum amount the firm should pay for this improvement?

Solution. Let X denote the maximum justifiable payment to improve the asset. The value of X can be found by equating the annual cost of the improved asset to that of the unimproved asset.

Under the original conditions,

$$\text{Annual cost} = 76,000R_9'' + 4,000(0.10) + 6,800$$
$$= 76,000(0.17364) + 4,000(0.10) + 6,800$$
$$= \$20,397$$

Equating the annual cost of the improved asset to this amount and selecting the installation date as the valuation date, we obtain the following values with respect to the improved asset:

$$\text{Value of payments} = 20,397\ {}_{11}t_0 = 20,397(6.49506)$$
$$= \$132,480$$

The payments under the revised conditions are recorded in Fig. 7-2. Evaluating all sums at the installation date and applying the value just calculated,

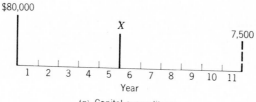

(a) Capital expenditures

(b) Annual maintenance

FIG. 7-2

we obtain the following:

$$80,000 + X(1.10)^{-5} - 7,500(1.10)^{-11} + 5,000\ {}_{11}t_0 + 1,800\ {}_5t_0 = \$132,480$$

Then
$$X(1.10)^{-5} = 15,810$$
$$X = 15,810(1.10)^5 = \$25,462$$

7-8. Amortization Method of Calculating Annual Cost. In Art. 7-6, we presented a derivation of Eq. (7-2) for the annual cost of an asset based on the assumption that purchase of the asset is financed with borrowed money, with interest on the loan to be paid annually and the principal to be repaid at the date of retirement of the asset. The assumption was also made that in order to accumulate this principal the firm made annual contributions to a sinking fund, whose interest rate equals the interest rate i of the loan. This fiscal arrangement was merely a hypothetical one, devised for convenience in establishing some basis for calculating annual cost, albeit an erroneous one.

The annual cost of an asset is sometimes calculated as though this method of financing were real rather than imaginary, with two distinct interest rates used for the loan and the sinking fund. This method also defines annual cost as the total of the disbursements made annually with regard to the asset.

Let i denote the interest rate of the loan and i' the interest rate of the sinking fund. The end-of-year disbursements attributable to the asset are the following:

1. A deposit in the sinking fund of

$$(P - L) \frac{i'}{(1 + i')^n - 1}$$

2. An interest payment of Pi
3. The annual operating cost O.C.

Totaling these, we obtain

$$\text{Annual cost} = (P - L) \frac{i'}{(1 + i')^n - 1} + Pi + \text{O.C.} \qquad (7\text{-}4)$$

The above procedure is known as the "amortization method" of calculating annual cost. While this method eliminates one of the three objections we have raised to the borrowed-money method, it fails to remove the others. Moreover, an additional objection to the amortization method arises from the fact that the establishment of a sinking fund is improbable, unless the borrowed money is obtained through the issuance of sinking-fund bonds.

Assuming that the asset is purchased with borrowed funds, some economists prefer to regard annual cost as the summation of the following expenses:

1. The depreciation charge for the year
2. The annual interest charge Pi
3. The annual operating cost O.C.

The annual cost obtained on this basis will depend on the method employed for allocating depreciation. If the sinking-fund method is used, then the annual deposit in the fund is represented by the first term of Eq. (7-4). From this point of view, the amortization method of calculating annual cost is also referred to as the "sinking-fund-depreciation method." There is an error inherent in this conception, however, for the actual depreciation charge made annually is the total of the deposit made in the hypothetical fund and the interest earned by this fund, the latter being a variable quantity.

A limitation of the depreciation method of viewing annual cost is that it is inadequate to cope with a situation where intermediate expenses of a nonuniform nature occur.

Example 7-9. (P.E. examination problem.) On the basis of the following data, which penstock would you choose?

	Material	
	Wood	Steel
First cost....................	$100,000	$150,000
Annual maintenance...........	$800	$500
Life in years.................	25	50

Taxes are 1 per cent of first cost, interest on loan is 6 per cent, and interest on reserve fund is 4 per cent compounded annually.

Solution

$$\text{Annual cost of wood penstock} = 100,000 \left[\frac{0.04}{(1.04)^{25} - 1} \right] + 100,000(0.06)$$
$$+ 800 + 1,000$$
$$= 2,401 + 6,000 + 800 + 1,000$$
$$= \$10,201$$

$$\text{Annual cost of steel penstock} = 150,000 \left[\frac{0.04}{(1.04)^{50} - 1} \right] + 150,000(0.06)$$
$$+ 500 + 1,500$$
$$= 983 + 9,000 + 500 + 1,500$$
$$= \$11,983$$

If this method of calculating annual cost is employed, the wood penstock is the economic choice.

7-9. Approximate Method of Calculating Annual Cost. In order to simplify the determination of annual cost, an approximate method is sometimes applied, which can be derived in the following manner: Assume that the funds required to purchase the asset are obtained by two loans, the first for the sum of $(P - L)$ and the second for the sum of L. The interest rate for both loans is i. The first loan requires that the principal be repaid in n equal annual installments of $(P - L)/n$ each, with interest to be paid annually on the outstanding principal. The second loan stipulates that the principal L is to be repaid as a lump sum at the end of n years, with interest paid annually. (Since the salvage value of the asset will provide the funds necessary to repay the principal, the only expense associated with this loan is the annual interest charge.) The annual payments required under this arrangement are the following:

1. A principal payment on the first loan of $(P - L)/n$
2. An interest payment on the second loan of Li
3. An interest payment on the first loan

The third item in this list of payments is a variable amount, whose value for any year can be determined as follows:

Year	Principal at beginning	Interest charge
1	$P - L$	$(P - L)i$
2	$(P - L)\dfrac{n - 1}{n}$	$(P - L)\dfrac{n - 1}{n}i$
3	$(P - L)\dfrac{n - 2}{n}$	$(P - L)\dfrac{n - 2}{n}i$
...	....	....
n	$(P - L)\dfrac{1}{n}$	$(P - L)\dfrac{1}{n}i$

To determine the annual cost, the time value of money is disregarded, or, to express this in another manner, the prevailing investment rate is taken as zero. Consequently, the time at which a disbursement is made is of no significance. In order to allocate the cost of the asset uniformly over its years of life, it is simply necessary to extract the arithmetical average of the total payments made. The first two payments tabulated above remain constant, but the interest payments on the first loan vary uniformly from $(P - L)i$ to $(P - L)i/n$. The average of these payments is

$$\frac{(P - L)i + (P - L)i/n}{2} = \frac{(P - L)i}{2}\left(1 + \frac{1}{n}\right)$$
$$= \frac{(P - L)i(n + 1)}{2n}$$

Hence,

$$\text{Annual cost} = \frac{P - L}{n} + \frac{(P - L)i(n + 1)}{2n} + Li + \text{O.C.} \quad (7\text{-}5)$$

It was stated above that annual cost is sometimes interpreted as comprising the annual depreciation charge, annual interest on first cost, and annual operating cost. If the straight-line method of allocating depreciation is used, the annual depreciation is represented by the first term in Eq. (7-5). The sum of the second and third terms of this equation represents average interest on first cost. Therefore, this approximate method of calculating annual cost is also referred to as the method of "straight-line depreciation plus average interest."

On the one hand, this method considers the time value of money by including the interest expense on borrowed money; on the other hand, it disregards the time value of money in allocating the invested capital to each year of the asset's life. Thus, it recognizes interest when it exists as an expense but does not when it exists in the form of potential income. Because the results obtained by application of this method are highly distorted, they are of little or no value, and there is scant justification for the use of this method.

The only accurate method of calculating the annual cost of an asset is the original one presented in Art. 7-3, in which all payments relating to the asset are converted to an equivalent annuity, and the interest rate used is the maximum rate obtainable by the organization owning the asset.

Example 7-10. A machine is to be purchased for a manufacturing plant, and two types of machine are under consideration. The estimated cost data pertaining to each are as follows:

	Machine A	Machine B
Purchase price...........	$6,000	$4,000
Salvage value............	$1,000	......
Life in years.............	12	10
Operating cost..........	$500	$650

If money can be borrowed at 6 per cent, which machine is more economical? Use straight-line depreciation and average interest cost.

Solution

Machine A:

$$\text{Annual cost} = \frac{5,000}{12} + \frac{5,000 \times 0.06 \times 13}{2 \times 12} + 1,000(0.06) + 500$$

$$= 417 + 163 + 60 + 500$$

$$= \$1,140$$

Machine B:

$$\text{Annual cost} = \frac{4,000}{10} + \frac{4,000 \times 0.06 \times 11}{2 \times 10} + 650$$

$$= 400 + 132 + 650$$

$$= \$1,182$$

By the use of this approximate method, machine A appears to be the economic choice.

7-10. Present Worth of Costs. We have found that the calculation of equivalent uniform annual cost constitutes a suitable technique for comparing the costs of alternative proposals. We shall now consider another method of cost comparison that is often used.

Assume that a machine is to be purchased to perform a certain manufacturing operation. Selecting one proposed type of machine, let us make the following set of calculations:

1. Convert the monetary values of all expenditures ascribable to this asset to their equivalent values as of the date of purchase. In making these conversions, we apply the same interest rate as we would in calculating annual cost. Salvage value is treated as a negative expenditure.

2. Total these equivalent values. The sum thus obtained is called the "present worth" of the costs of the asset.

Now assume that two alternative types of machine are available. If the machines have identical life spans, their costs can be compared directly by calculating the present worth of costs of each. Generally, however, the alternative machines will have unequal life spans. By modifying the procedure for calculating present worth of costs, it is possible to extend this method of cost comparison to the general as well as the special case.

In our study of annual cost, we assumed that the history of one life of an asset will be duplicated in each subsequent life, ad infinitum. Thus, the process of retiring one asset and replacing it with a new one can be viewed as simply a renewal of the original asset, which is considered to remain in service perpetually. Therefore, having the cost data pertaining to one life of an asset, we can calculate the present worth of costs for several lives, the "present" being the date at which the first life begins. Thus, if two alternative assets have unequal life spans, their costs can be compared by calculating the present worth of costs for a time period that is an exact multiple of the life span of each asset. This time period is referred to as the "analysis period."

Example 7-11. Make a cost comparison of the following assets (*a*) by the annual-cost method; (*b*) by the present-worth method. Use a 7 per cent interest rate.

	Machine A	Machine B
First cost...............	$12,000	$13,600
Salvage value...........	$1,500	....
Life in years............	6	8
Operating cost..........	$900	$700

Solution. *Part a.* We shall apply Eq. (7-1*a*).

Machine A:

$$\text{Annual cost} = 10{,}500R_6'' + 1{,}500(0.07) + 900$$
$$= 10{,}500(0.20980) + 1{,}500(0.07) + 900$$
$$= 2{,}203 + 105 + 900$$
$$= \$3{,}208$$

Machine B:

$$\text{Annual cost} = 13{,}600R_8'' + 700$$
$$= 13{,}600(0.16747) + 700 = 2{,}278 + 700$$
$$= \$2{,}978$$

Part b. We select an analysis period of 24 years, since 24 is the lowest common multiple of 6 and 8. Thus, we are dividing time into 24-year cycles and analyzing the events of each cycle.

The capital expenditures for machine A during the first 24-year cycle are shown in Fig. 7-3. At the end of the first, second, and third lives, the asset is renewed at a cost of $12,000 − $1,500, or $10,500. At the end of the fourth life,

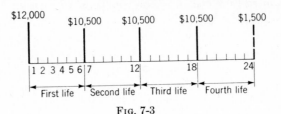

Fig. 7-3

the asset is scrapped, and an income of $1,500 is realized. (The payment of $12,000 at the end of the fourth life to renew the asset belongs to the second 24-year cycle.)

Machine A:

$$\text{Present worth} = 12{,}000 + 10{,}500[(1.07)^{-6} + (1.07)^{-12} + (1.07)^{-18}] \\ - 1{,}500(1.07)^{-24} + 900 \, _{24}t_0$$

The calculations can be tabulated in the manner shown in Table 7-1.

TABLE 7-1. PRESENT WORTH OF COSTS, MACHINE A

First cost	$12,000
Renewal cost:	
$(1.07)^{-6} = 0.66634$	
$(1.07)^{-12} = 0.44401$	
$(1.07)^{-18} = 0.29586$	
$\overline{1.40621} \times 10{,}500$	= 14,765
Operating cost:	
$900 \, _{24}t_0 = 900(11.46933)$	= 10,322
Total	= $37,087
Salvage value:	
$1{,}500(1.07)^{-24} = 1{,}500(0.19715) =$	296
Present worth of costs	= $36,791

Alternatively, the total renewal cost for the analysis period can be found by considering that the renewal payments of $10,500 each constitute an annuity having a 6-year payment period. The renewal cost is then calculated by applying Eq. (4-24).

Machine B: The calculations are shown in Table 7-2.

TABLE 7-2. PRESENT WORTH OF COSTS, MACHINE B

First cost and renewal cost:

$$1.00000$$
$$(1.07)^{-8} = 0.58201$$
$$(1.07)^{-16} = \underline{0.33873}$$
$$\overline{1.92074} \times 13,600 = \$26,122$$

Operating cost:

$$700 \; _{24}t_0 = 700(11.46933) \qquad = \underline{8,029}$$
$$\text{Present worth of costs} \qquad = \$34,151$$

Summary. The cost comparison reveals that machine B is more economical than machine A. The results obtained by the annual-cost and present-worth methods can be tested for compatibility by calculating the ratio of the cost of the assets.

$$\text{Annual-cost ratio} = \frac{3{,}208}{2{,}978} = 1.077$$

$$\text{Present-worth ratio} = \frac{36{,}791}{34{,}151} = 1.077$$

Both methods reveal that the cost of machine A is 7.7 per cent higher than that of machine B.

7-11. Capitalized Cost. In making a cost comparison of two alternative assets by the present-worth method, we compute the present worth of costs for an analysis period that comprises an integral number of lives of each asset. The calculations can be simplified by selecting an analysis period that comprises an infinite number of lives. The present worth of costs associated with an asset for an infinite period of time is referred to as its "capitalized cost." This value can be regarded as the amount of capital which, if invested in a fund earning a stipulated interest rate, will be just sufficient to provide all payments required to maintain the asset in perpetual service. However, it must be emphasized that the interest rate to be applied in calculating capitalized cost is the highest rate of return the firm can obtain through some alternative investment.   Capitalized cost, like annual cost, is a relative rather than an absolute quantity; it is calculated as a means of comparing two alternative proposals.

Since our summation of the costs ascribable to an asset applies to an infinity of lives, each disbursement will recur after a specific interval of time and continue to recur indefinitely.   For example, if a machine costs $10,000 and has a life span of 8 years and a salvage value of $500, then the payment required to renew the original asset is $9,500, to be made 8 years after the original purchase date.   This renewal cost of $9,500 will recur at the end of each 8-year period indefinitely.   Hence, the renewal costs constitute a perpetuity, whose origin date is the original purchase date of the asset.

By Eq. (4-27), if a perpetuity consists of payments of R each, made at intervals of n interest periods, and i is the interest rate, the value of the perpetuity at its origin date is

$$M = \frac{R}{(1 + i)^n - 1}$$

Continuing the previous notation, let P denote the first cost of the asset, L its salvage value, and n the number of years in its life span.   If we assume there are no intermediate payments required other than the annual operating costs and that these remain constant, the expenditures associated with the asset can be classified as follows:

1. The first cost P
2. The periodic renewal cost of $(P - L)$, recurring at intervals of n years; by Eq. (4-27), the value of this perpetuity at the original purchase date is

$$M = \frac{P - L}{(1 + i)^n - 1}$$

3. The annual operating costs; the value of this perpetuity is, by Eq. (4-27),

$$M = \frac{\text{O.C.}}{i}$$

Totaling the values of these three groups, we obtain

$$\text{Capitalized cost} = P + \frac{P - L}{(1 + i)^n - 1} + \frac{\text{O.C.}}{i} \qquad (7\text{-}6)$$

This equation can be derived by an alternative method.   In Art. 7-6, we converted the expenses incurred in the use of an asset to an equivalent series of uniform annual payments, the amount of each payment being designated as the annual cost of the asset. Since these payments are assumed to continue indefinitely, they constitute a perpetuity.   Hence, applying Eq. (4-27) to the special case where the interval between payments is one interest period, we have

$$\text{Capitalized cost} = \frac{\text{annual cost}}{\text{interest rate}}$$

If we substitute the expression for annual cost as given by Eq. (7-2), we obtain Eq. (7-6) above.   However, by applying Eq. (7-1) for annual cost we obtain an alternative equation for capitalized cost, namely,

$$\text{Capitalized cost} = \frac{P - L}{1 - (1 + i)^{-n}} + L + \frac{\text{O.C.}}{i} \qquad (7\text{-}7)$$

In order to obviate the necessity of dividing by an odd figure, we shall transform the fractions having $(P - L)$ in the numerator into a simpler form by dividing both numerator and denominator by i.   This procedure yields

$$\frac{P - L}{(1 + i)^n - 1} = \frac{(P - L)/i}{[(1 + i)^n - 1]/i} = \frac{P - L}{i} \frac{i}{(1 + i)^n - 1}$$
$$= \frac{P - L}{i} R'_n$$

Similarly,

$$\frac{P - L}{1 - (1 + i)^{-n}} = \frac{P - L}{i} R''_n$$

Substituting in the above equations for capitalized cost, we obtain

$$\text{Capitalized cost} = P + \frac{P - L}{i} R'_n + \frac{\text{O.C.}}{i} \qquad (7\text{-}6a)$$

and $$\text{Capitalized cost} = \frac{P - L}{i} R''_n + L + \frac{\text{O.C.}}{i} \qquad (7\text{-}7a)$$

Where expenses of a nonuniform nature occur in each life of the asset, it is simply necessary to augment the capitalized cost as given by either of these two equations to include the perpetuity formed by the non-uniform payments.   Caution must be exercised, however, to establish correctly the origin date of the perpetuity. Thus, assume that the life span of an asset is 8 years and that major repairs are required at the end of the fifth year. This situation is represented diagrammatically in Fig. 7-4. The expenditures for these repairs in each life of the asset constitute a perpetuity whose payment period is 8 years. Hence, the origin date of this perpetuity is 8 years prior to the first payment, or 3 years prior to the original purchase date. To obtain the capitalized cost of the asset, we can evaluate the perpetuity at its origin date and then translate this value to its equivalent value at the purchase date.

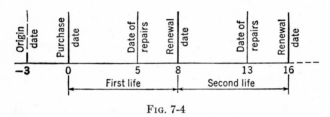

FIG. 7-4

If an asset is considered to have an infinite life span, the renewal cost is zero, and therefore

$$\text{Capitalized cost for perpetual life} = P + \frac{\text{O.C.}}{i} \qquad (7\text{-}8)$$

Example 7-12. (P.E. examination problem.) Two methods, A and B, of conveying water are being studied. Method A requires a tunnel: first cost $150,000, life perpetual, annual operation and upkeep $2,000. Method B requires a ditch plus a flume: first cost of the ditch is $50,000, life perpetual, annual operation and upkeep $2,000; first cost of flume is $30,000, life 10 years, salvage value $5,000, annual operation and upkeep $4,000. Compare the two methods for perpetual service. The interest rate is 5 per cent.

	Method A, tunnel	Method B	
		Ditch	Flume
First cost..................	$150,000	$50,000	$30,000
Salvage value..............	....	....	$5,000
Life in years..............	Infinite	Infinite	10
Operating cost..............	$2,000	$2,000	$4,000

Solution

Method A:

$$\text{Capitalized cost of tunnel} = 150{,}000 + \frac{2{,}000}{0.05}$$
$$= 150{,}000 + 40{,}000$$
$$= \$190{,}000$$

Method B:

$$\text{Capitalized cost of ditch} = 50{,}000 + \frac{2{,}000}{0.05}$$
$$= 50{,}000 + 40{,}000$$
$$= \$90{,}000$$
$$\text{Capitalized cost of flume} = 30{,}000 + \frac{25{,}000}{0.05}R'_{10} + \frac{4{,}000}{0.05}$$
$$= 30{,}000 + 500{,}000(0.07950) + 80{,}000$$
$$= \$149{,}750$$
$$\text{Total for method B} = \$239{,}750$$

Method A is the economic choice.

Example 7-13. (P.E. examination problem.) Compare two types of bridges, A and B, on the basis of capitalized cost at 5 per cent interest. Bridge A has an estimated life of 25 years, initial cost of $50,000, renewal cost of $35,000, annual maintenance of $500, repairs every 5 years amounting to $2,000, and salvage value of $5,000. Bridge B has an estimated life of 50 years, initial cost of $75,000, renewal cost of $75,000, annual maintenance cost of $100, repairs every 5 years amounting to $1,000, and salvage value of $10,000. The initial cost can be paid out of available funds. All other expenses will be defrayed by sinking funds.

Solution. The final two sentences have no significance in the calculation of capitalized cost. With regard to bridge A, the net renewal cost is $35,000 minus $5,000, or $30,000. In both instances, repairs are made at 5-year intervals. These payments, however, do not form a continuous series, for no repairs are made at the date of renewal. To circumvent this obstacle, we shall assume that disbursements for repairs are actually made at the renewal dates and then deduct the value of these imaginary payments. By extending Eq. (7-6) to include the expenditures for repairs, we obtain

$$\text{Capitalized cost of bridge A} = 50{,}000 + \frac{30{,}000}{(1.05)^{25} - 1} + \frac{500}{0.05} + \frac{2{,}000}{(1.05)^{5} - 1}$$
$$- \frac{2{,}000}{(1.05)^{25} - 1}$$

We can avoid dividing by an odd amount if we divide both numerator and denominator of the second, fourth, and fifth terms by 0.05. This yields

$$\text{Capitalized cost} = 50,000 + \frac{30,000}{0.05} R'_{25} + 10,000 + \frac{2,000}{0.05} R'_5 - \frac{2,000}{0.05} R'_{25}$$

$$= \$78,971$$

$$\text{Capitalized cost of bridge B} = 75,000 + \frac{65,000}{(1.05)^{50} - 1} + \frac{100}{0.05} + \frac{1,000}{(1.05)^5 - 1}$$

$$- \frac{1,000}{(1.05)^{50} - 1}$$

$$= 75,000 + \frac{65,000}{0.05} R'_{50} + \frac{100}{0.05} + \frac{1,000}{0.05} R'_5$$

$$- \frac{1,000}{0.05} R'_{50}$$

$$= \$86,738$$

Bridge A is the economic choice.

7-12. Economical Retirement Date. In our preceding study of the costs associated with the purchase, operation, and replacement of an asset, we have assumed that the annual cost of operation, which includes maintenance, periodic repairs, etc., remains constant during the entire life of the asset. In reality, however, such is invariably not the case, as the annual cost of operating a particular asset tends to increase with the age of the asset. Moreover, the problems treated in the preceding sections considered the longevity of the asset to be a fixed period of time, which is also a somewhat unrealistic conception. An asset does not abruptly expire; it ages and gradually reaches a state where its operating cost is excessively high compared to the monetary returns derived from its use, and the retirement of the asset then becomes feasible. For this reason, there is a fundamental problem applying to every operating asset, namely, to determine the precise date at which it is most economical to retire the asset. We shall assume, as previously, that exact cost data are available.

While the problem of the most economical retirement date for an asset has many ramifications, the basic method of approach is identical for all. We shall consider the situation in which an asset, upon its retirement, is to be replaced by a duplicate asset; this process of retirement and replacement is presumed to continue indefinitely.

There are three factors that determine the average annual cost corresponding to a particular life span of an asset: namely, the purchase price or first cost; the total operating costs incurred during that life span; and the salvage value procurable for the old asset, either in the form of an allowance toward the purchase of its replacement or through the sale of the old asset as scrap. To allow us to focus our attention more sharply upon these factors, we shall disregard momentarily the time value of money, thus permitting the addition and subtraction of sums of money without regard to their corresponding dates.

Let P denote the first cost of the asset, L_n the salvage value at the end of the nth year, and Z the total operating cost up to the end of the nth year. Then, for a life span of n years, we have

$$\text{Average annual cost of asset} = \frac{P - L_n + Z}{n} = \frac{P - L_n}{n} + \frac{Z}{n}$$

$$= \text{average annual depreciation}$$
$$+ \text{average annual operating cost}$$

As the age of the asset increases, not only does the average annual operating cost increase continually, but it increases at an accelerated pace. On the other hand, the average annual depreciation varies in precisely the opposite manner, for the salvage value declines very rapidly during the asset's first year and thereafter continues to decline at a decelerated pace. There are several reasons contributing to this pattern of depreciation. When an asset is purchased, there are certain fixed costs associated with transporting it to the site and assembling it for operation; these fixed expenses, being irretrievable, immediately become one of the constituents of depreciation, irrespective of the asset's longevity. Similarly, in the retirement of the asset, dismantling and relocating it incur other fixed costs that represent a permanent impairment of the salvage value. Furthermore, aside from the question of obsolescence due to technological improvements or change of style (which we are excluding from consideration), there is the inevitable loss in the market value of an asset once it has been purchased and has acquired the appellation "secondhand."

While the average annual depreciation and average annual operating cost exert contrary effects upon the average annual cost of the asset, these two forces are largely unrelated to one another. The result is that there is a distinct average annual cost corresponding to every possible life span of the asset.   The particular life span for which the annual cost is minimum establishes the date at which it is most economical to retire the asset.

The determination of this retirement date involves a choice among a number of possibilities, since the asset can be retired at the end of the first, second, third, etc., years of its life.   We shall assume again that each cycle of events will be repeated in unbroken sequence.   Thus, if the original asset is to be retired at the end of the fifth year, each succeeding asset will also be retired at the end of the fifth year of its life span. Although for simplicity we disregarded the time value of money in our cursory analysis above, we must of course include this factor in our determination of cost.   The costs associated with each of the alternative life spans can be compared on the basis of either equivalent uniform annual cost or capitalized cost.   We shall apply the former method.

The calculation of the annual cost of an asset corresponding to each life span can be performed in the manner described above, that is, by taking all sums of money associated with that life span and converting them, either singly or in combination, to an equivalent annuity.   This direct method, however, is quite laborious because it involves extensive multiplication of odd amounts.   A simpler method of calculating the annual cost can be obtained in the following manner: By Eq. (7-2a),

$$\text{Annual cost} = (P - L)R'_n + Pi + \text{O.C.}$$

In order to apply this equation to our present problem of economical retirement, it becomes necessary to modify it somewhat to take into account the fact that the operating costs do not remain constant but vary from year to year.   (The salvage value L is also a function of the life span.)   The first term of this equation represents the cost of replacing the asset, prorated uniformly over the years of the asset's life.   It can thus be regarded as the equivalent uniform annual replacement cost. The second term of this equation represents potential income that is annually forfeited through the investment of capital in this asset.   Since this quantity is independent of the life span and since we are only interested in the differences of annual cost, we shall exclude this from our calculations.   Finally, for the third term of Eq. (7-2a), we can substitute a quantity that represents an equivalent uniform annual operating cost. Let $C_1, C_2, C_3, \ldots C_n$ denote the operating costs at the end of the first, second, third, $\ldots$ nth years, respectively, and Q_n denote the equivalent uniform annual operating cost for a life span of n years. Then, by definition,

$$Q_n = [C_1(1 + i)^{-1} + C_2(1 + i)^{-2} + \cdots + C_n(1 + i)^{-n}]R_n''$$

or $\quad Q_n = [C_1(1 + i)^{n-1} + C_2(1 + i)^{n-2} + \cdots + C_n]R_n'$

Assume that the equivalent uniform annual operating cost Q_n at the end of the nth year has been determined. The corresponding cost Q_{n+1} at the end of the $(n + 1)$st year can be easily obtained by applying the following relationship:

$$Q_{n+1} = Q_n + (C_{n+1} - Q_n)R_{n+1}' \tag{7-9}$$

Proof. In Fig. 7-5a, we have represented the disbursements for the annual operating costs of C_1, C_2, and C_3 associated with a life span of 3 years. The sequence of payments is assumed to recur in a fixed pattern

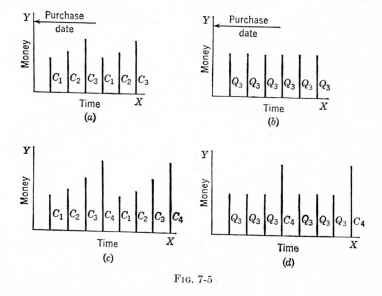

FIG. 7-5

for an indefinite period of time. Since Q_3 represents a uniform annual payment that is equivalent to the three variable payments, we can replace the actual payments with this constant theoretical payment, as shown in Fig. 7-5b.

Similarly, the payments C_1, C_2, C_3, and C_4 pertaining to a life span of 4 years are represented in Fig. 7-5c. We can again replace the first three payments with their uniform equivalent Q_3, as shown in Fig. 7-5d. In order to equalize the four payments, it is now simply necessary to find the excess of C_4 over Q_3 and to allocate this excess uniformly over the 4 years of the asset's life, through conversion to an equivalent annuity. Since this excess payment $(C_4 - Q_3)$ occurs at the end of the fourth year, the annual payment of its equivalent annuity is

$$(C_4 - Q_3)R_4'$$

This amount must be added to Q_3 to obtain the equivalent uniform annual payment Q_4 corresponding to a life span of 4 years.

Although in the above discussion we have assigned a particular value to the life span n, the reasoning is perfectly general and applies to all values of n. Equation (7-9) is thus established. While we have derived this equation on a purely logical basis, it can also be derived mathematically.

Consider now the case where n is 1; that is, the asset is retired and replaced at the end of each year. Under this arrangement, the annual cost of operating the asset remains constant at C_1; hence, $Q_1 = C_1$. Since Q_1 is known, Q_2 can be easily calculated by applying Eq. (7-9); with Q_2 determined, we proceed to calculate Q_3 in a similar manner, etc. Thus, starting with the known quantity Q_1, we can obtain Q_n for any value of n through the process of successively applying Eq. (7-9).

Summarizing the above results, we have

Net annual cost = equivalent uniform annual replacement cost
$\qquad\qquad$ + equivalent uniform annual operating cost
$$= (P - L_n)R_n' + Q_n \qquad\qquad (7\text{-}10)$$

We have designated this quantity as "net annual cost" rather than annual cost by reason of the fact that the constant Pi (interest on first cost) is not included in this quantity.

Example 7-14. A machine was purchased and installed at a total cost of $10,000. On the basis of the following data, at what date should the machine be retired? Use an annual interest rate of 7 per cent.

COST COMPARISON OF ALTERNATIVE METHODS 177

End of year	Salvage value	Annual operating cost
1	$6,000	$ 2,300
2	4,000	2,500
3	3,200	3,300
4	2,500	4,800
5	2,000	6,800
6	1,800	9,500
7	1,700	12,000

Solution. This problem is solved in Tables 7-3, 7-4, and 7-5. The machine should be retired at the end of the third year.

TABLE 7-3. EQUIVALENT UNIFORM ANNUAL REPLACEMENT COST

Year	$(P - L)R_n'$	
1	4,000(1)	= $4,000
2	6,000(0.48309) =	2,899
3	6,800(0.31105) =	2,115
4	7,500(0.22523) =	1,689
5	8,000(0.17389) =	1,391
6	8,200(0.13980) =	1,146
7	8,300(0.11555) =	959

TABLE 7-4. EQUIVALENT UNIFORM ANNUAL OPERATING COST

Year	Calculated by Eq. (7-9)	
1	$Q_1 = 2,300$	
2	$Q_2 = 2,300 + 200(0.48309)$	= $2,397
3	$Q_3 = 2,397 + 903(0.31105)$	= 2,678
4	$Q_4 = 2,678 + 2,122(0.22523)$	= 3,156
5	$Q_5 = 3,156 + 3,644(0.17389)$	= 3,790
6	$Q_6 = 3,790 + 5,710(0.13980)$	= 4,588
7	$Q_7 = 4,588 + 7,412(0.11555)$	= 5,444

TABLE 7-5. NET ANNUAL COST

Year		
1	4,000 + 2,300 =	$6,300
2	2,899 + 2,397 =	5,296
3	2,115 + 2,678 =	4,793
4	1,689 + 3,156 =	4,845
5	1,391 + 3,790 =	5,181
6	1,146 + 4,588 =	5,734
7	959 + 5,444 =	6,403

7-13. The Rational Method of Solving Problems of Finance. The present chapter concludes that phase of our study of finance that is concerned with the time value of money. It is therefore appropriate at this point to make the following observation: Many problems of finance are susceptible of solution either on a wholly mathematical basis or on a strictly logical basis involving only common-sense financial considerations. For problems of this nature, it is generally preferable to apply the logical rather than the mathematical method of solution, for elementary logic alone can often disclose simple relationships that are obscured by intricate mathematical formulas. This method has been demonstrated in this text on several previous occasions, as in the alternative derivation of the equation for the sinking-fund principal and in the derivation of the equation for the equivalent uniform annual operating cost. However, the application of the logical, or rational, method will now be illustrated again to emphasize its importance.

Assume that we wish to make a ratio comparison of the quantity R'_n, the periodic deposit required to develop a principal of \$1 in a sinking fund at the end of the nth period, based on an interest rate i, with its successor R'_{n+1}, the periodic deposit required to develop this principal in $(n + 1)$ periods. Now, consider the sum R'_n to be deposited in a sinking fund, not n times but $(n + 1)$ times, as represented in Fig. 7-6. At the end of the nth period, the principal in the fund is \$1; at the end of the $(n + 1)$st period, through the accrual of interest and the receipt of an additional

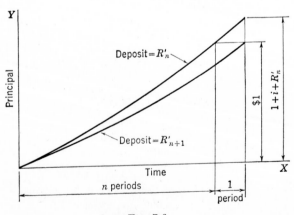

Fig. 7-6

deposit, the principal expands to $(1 + i + R'_n)$. On the other hand, if the sum R'_{n+1} is deposited $(n + 1)$ times, the principal at the end of the $(n + 1)$st period is simply \$1. Since the principal in the fund at a given date is directly proportional to the periodic deposit, we obtain the ratio

$$\frac{R'_n}{R'_{n+1}} = \frac{1 + i + R'_n}{1} = 1 + i + R'_n$$

Applying Eq. (4-6), this ratio can be expressed as

$$\frac{R'_n}{R'_{n+1}} = 1 + R''_n$$

Similarly, if we compare the respective principals at the end of the nth period, we obtain

$$\frac{R'_n}{R'_{n+1}} = \frac{1 + i}{1 - R'_{n+1}}$$

The logical method can likewise be applied to derive Eq. (4-6),

$$R'_n + i = R''_n$$

which we previously derived mathematically. Assume that two sinking funds are established simultaneously, each earning interest at the identical rate i and each having a term of n periods. The periodic deposit in the first fund is R'_n, while that in the second fund is R''_n. By definition of the quantities R'_n and R''_n, the principal in the first fund is \$1 at its terminal date, and the second fund has a value of \$1 at its origin date. Hence, by the application of the basic money-time relationship, the principal in the second fund at its terminal date is $(1 + i)^n$. Let M denote the difference between the principals in the two funds at the terminal date. Then

$$M = \text{Principal in second fund} - \text{principal in first fund}$$

or $\quad M = (1 + i)^n - 1$

However, by applying the sinking-fund principal equation, we obtain the following:

$$M = (R''_n - R'_n)\frac{(1 + i)^n - 1}{i}$$

Equating these two expressions for M produces

$$R_n'' - R_n' = i$$

or

$$R_n' + i = R_n''$$

PROBLEMS

7-1. An asset had an initial cost of $6,000, and it was retired at the end of 7 years with a salvage value of $400. The cost of operation and maintenance was $550 annually. What was the annual cost of this asset, based on an interest rate of 6½ per cent? Verify the solution by calculating the effective purchase price at date of purchase and converting this sum to an equivalent annuity.

7-2. (P.E. examination problem.) Assume you are required to decide on which of two machines to purchase for a given manufacturing process. Both machines have been judged to be equally desirable in all matters except cost. The following data are available:

	Machine A	Machine B
Delivered cost....................	$25,000	$40,000
Installation expense..............	$5,000	$10,000
Estimated life in years...........	6	8
Net salvage value................	None	$10,000

	Machine A	Machine B
Annual operating costs:		
Labor.........................	$3,000	$2,000
Power.........................	2,000	1,750
Repairs.......................	2,000	1,000
Taxes.........................	450	750
Interest Rate....................	6%	6%

Which machine will have the lower annual cost and what annual savings will accrue from its purchase?

7-3. (P.E. examination problem.) A company requires 800,000 kwhr of electric energy per year, which can be purchased at 1.3 cents per kilowatthour. Will it pay the company to build a power plant of the above capacity under the following assumptions: first cost, $60,000; annual operation and maintenance, $7,500; life, 15 years; salvage value, $5,000; insurance, 1 per cent; cost of money, 4 per cent?

Ans. Annual cost of power plant = $13,247; annual cost of purchasing energy = $10,400

7-4. (P.E. examination problem.) A snow-loading machine costing $10,000 requires four operators at $12 per day. The machine can do the work of 50 hand shovelers at $7 per day. Fuel, oil, and maintenance for the machine amount to $30 per day. If the life of the machine is 8 years and the interest on money is 6 per cent, how many days of snow removal per year are necessary to make purchase of the machine economical?

Solution. Let x = number of days of snow removal per year.

$$\text{Annual cost of machine} = \$1,610 + 78x$$
$$\text{Annual cost of snow removal without machine} = 350x$$

Therefore 6 days of snow removal annually are required to warrant purchase of machine.

7-5. An asset was purchased for $8,500, and it was retired at the end of 8 years with a salvage value of $700. Its annual maintenance cost was $600 for the first 3 years and $900 for the remaining 5 years. What was the annual cost of this asset if money is worth 7 per cent?

Solution

$$\text{Equivalent uniform annual maintenance cost} = \$768$$
$$\text{Annual cost} = \$2,123$$

7-6. On the basis of the following data, calculate the annual cost of an asset:

$$\text{First cost} = \$30,000$$
$$\text{Life} = 21 \text{ years}$$
$$\text{Salvage value} = \$2,000$$
$$\text{Annual operating cost} = \$1,200$$
$$\text{Repairs every 3 years (except at the date of retirement)} = \$3,500$$

The firm owning the asset can obtain an 8 per cent return on its invested capital.

Ans. $5,164

7-7. An asset costing $20,000 will have a salvage value of $5,000 at the expiration of 8 years and of $1,000 at the expiration of 10 years. If we assume that the annual operating cost remains constant, which of the alternative life spans is preferable? Money is worth 7 per cent.

Ans. The annual cost corresponding to a 10-year life is $87 lower than that corresponding to an 8-year life.

7-8. A machine has a first cost of $30,000, and it will be retired at the end of 10 years with an anticipated salvage value of $4,000. The annual operating cost is $3,500. Calculate the annual cost of the machine by each of the following methods:

a. Amortization method. The interest rate of the loan is 6 per cent, and the interest rate of the sinking fund is 3 per cent.

b. Method of straight-line depreciation plus average interest. Use a 6 per cent rate.

c. Conversion of all payments to an equivalent annuity, based on a 6 per cent investment rate.

7-9. Two different types of equipment are being considered for performing a particular operation. Compare the two types on the basis of annual cost, using an interest rate of 6 per cent.

	Type A	Type B
First cost....................	$52,000	$30,000
Life in years................	12	6
Salvage value...............	$6,000	$2,000
Annual maintenance..........	$2,800	$3,050

Ans. Type A, $8,647; type B, $8,864

7-10. In the preceding problem, the manufacturer of the type B equipment estimates that the service life of the asset can be extended to 8 years if a new motor is installed at the end of the fourth year. This will not affect the salvage value or annual maintenance cost. What is the maximum allowable cost of the new motor to justify purchase of the type B equipment?   *Ans.* $7,589

7-11. Compare the following assets on the basis of present worth of costs, using an interest rate of 8 per cent. (Compute present worth to the nearest $100.)

	Machine A	Machine B
First cost...............	$60,000	$40,000
Salvage value...........	$5,000	$3,000
Life in years............	10	8

The estimated annual operating cost is as follows: for machine A, $4,000; for machine B, $5,000 for the first 2 years and $6,000 thereafter.

Ans. For a 40-year analysis period, present worth of costs is $150,200 for A and $147,500 for B.

7-12. A bridge that was constructed at a cost of $75,000 is expected to last 30 years, at the end of which time its renewal cost will be $40,000. Annual repairs and maintenance are $3,000. What is the capitalized cost of the bridge, at an interest rate of 6 per cent?   *Ans.* $133,433

7-13. It was stated in the text that the capitalized cost of an asset can be regarded as representing the sum of money which, if invested on the initial date at the stipulated interest rate, is just sufficient to provide for the purchase, maintenance, and perpetual renewal of the asset. Applying this conception, verify the solution to the preceding problem by tracing the history of a hypothetical fund in which the sum of $133,433 is deposited at a 6 per cent interest rate and from which all payments pertaining to the asset are withdrawn.

7-14. An asset was purchased for $100,000 and retired at the end of 15 years with a salvage value of $4,000. The annual operating cost was $18,000. Calculate the capitalized cost of the asset based on an interest rate of 8 per cent. *Ans.* $369,200

7-15. Compare the following two bridges on the basis of capitalized cost, with interest at 7 per cent.

	Bridge A	Bridge B
First cost....................	$100,000	$80,000
Life in years................	40	30
Renewal cost...............	$70,000	$55,000
Annual maintenance.........	$800	$1,300

Bridge A requires repairs of $2,000 at the expiration of every 8-year period, while bridge B requires repairs of $2,500 at the expiration of every 5-year period. (No repairs are made at the renewal date.)

Ans. Bridge A, $119,080; bridge B, $112,720

7-16. The operating costs incurred during the first 3 years of the life of an asset were as follows:

End of first year $2,000
End of second year   3,000
End of third year   4,500

Based on an interest rate of 5 per cent, calculate the equivalent uniform annual operating cost of this asset for the 3-year period under consideration with the following methods:

a. By evaluating each payment at the purchase date and converting the total amount to an equivalent annuity.

b. By applying Eq. (7-9).

7-17. Prove that Eq. (7-2a) can be derived from Eq. (7-1a) by applying Eq. (4-6),

$$R'_n + i = R''_n$$

7-18. (P.E. examination problem.) A full development of part of a certain water-supply project is to be compared with a stepped program of development. The full development involves a first cost of $1,400,000 for properties that are considered to be permanent. Annual operation and upkeep costs are $120,000. The stepped program involves an investment of $800,000 now and $1,000,000 at a date 15 years hence; both investments are for properties believed to be permanent. Annual operation and upkeep will be $100,000 for the first 15 years and $150,000 thereafter. With interest at 4 per cent, compare the capitalized costs of these alternatives.

7-19. (P.E. examination problem.) Plan A calls for an initial investment of $300,000 and expenditures of $10,000 a year for the first 20 years and $20,000 a year thereafter. It also calls for the expenditure of $200,000 at a date 20 years hence and every twentieth year thereafter.

Plan B calls for an initial investment of $500,000 followed by a single investment of $100,000 thirty (30) years hence. It also involves periodic expenditures of $50,000 every 10 years.

Compare plan A and plan B on the basis of the capitalized cost of perpetual service, using interest at 4 per cent.

7-20. Solve Example 7-8 by the capitalized-cost method.

Solution. Under the original conditions, capitalized cost of facility = \$203,966. Let M denote the value at a given date of all future payments for an infinite number of lives. At the end of the ninth year (when the second life begins), M = capitalized cost = \$203,966. At the end of the fifth year,

$$M = 203,966(1.10)^{-4} + 6,800 \, _4t_0 - 4,000(1.10)^{-4} = \$158,134$$

Assume for simplicity that the proposed improvement will be made only in the first life. Under the revised conditions, at the end of the fifth year,

$$M = 203,966(1.10)^{-6} + 5,000 \, _6t_0 - 7,500(1.10)^{-6} = \$132,675$$

Then
$$X = 158,134 - 132,675 = \$25,459$$

CHAPTER 8

ANALYSIS OF BUSINESS OPERATIONS

8-1. Introduction. The basic purpose of studying engineering economics is to determine either the manner in which a given sum of capital can be employed to secure a maximum rate of return or the manner in which a required task can be performed at minimum cost. Generally, there are multiple ways of investing the capital or multiple methods of performing the required task. We shall refer to each possible investment or each possible method of performance as a "scheme." Thus, there exists in engineering economics a wide class of problems in which it is necessary to make an economic choice among alternative schemes. The solution of these problems requires an analysis of the monetary returns or costs associated with each scheme.

In the preceding chapter, we formulated techniques of cost comparison where the payments corresponding to the alternative methods of performance differed from one another not only in amount but in time as well. In recognition of the time value of money, it was necessary to weave into our study the various formulas pertaining to compound interest. In this chapter, on the other hand, we shall investigate problems in which the payments corresponding to each alternative scheme are made simultaneously or are separated from one another by a time interval not exceeding 1 year. The time value of money has no bearing in the former case; in the latter situation, it enters into the problem clothed only in the form of simple interest.

8-2. Kelvin's Law. In many problems involving economic choice that arise in engineering economics and business management, a decision is required regarding the size or quantity most suitable for a particular asset or process. If we compare two different available sizes, we shall find in many instances that certain elements of cost are greater for the larger size, while other elements of cost are smaller for the larger size. Some typical examples follow.

Economical Span Length of a Bridge.   If a bridge is to be constructed to span a river, a decision is required concerning the number of spans into which the bridge is to be divided.   The greater the span length selected, the greater the weight of the steel required for the superstructure. On the other hand, the greater the span length, the fewer the piers to be constructed.

Economical Lot Size.   Assume that a stipulated number of units of a particular commodity are to be manufactured in the course of one year. This total quantity will be divided into a number of equal groups, or lots, each lot to be manufactured in one run.   The runs will be spaced at equal intervals of time.   Hence, it is necessary to determine the number of units to be assigned to each lot.

There are many fixed expenses incurred with each run, such as the cost of setting up the machines to be used.   It thus follows that the greater the lot size selected (and therefore the lower the number of runs required), the smaller the annual cost associated with the preparation for manufacture.   However, an increase in lot size produces an increase in such expenses as storage, insurance on inventory, and taxes.

There also enters into this problem the time value of money, in the form of simple interest.   To illustrate this, assume that 6,000 units are to be produced in 1 year and then consider the following two cases:

1. The units are to be produced in 2 lots of 3,000 each.   Production of the first lot is to commence January 1, and production of the second lot is to commence July 1.

2. The units are to be produced in 12 lots of 500 each, and production of a lot is to commence at the beginning of each month.

It is evident that the capital required to produce the 6,000 units is consumed more rapidly under the first scheme than under the second. Hence, the greater the lot size selected, the greater the loss to the business firm of potential income from invested capital.   In order to give weight to this factor, an item termed "interest expense" is often added to the cost for each particular lot size.   It is to be understood, however, that unless the money to finance the business enterprise is actually borrowed, there is, in the narrow sense of the term, no interest expense present. This item merely reflects the interest income which the firm has forfeited through investment of its capital in inventory.

Economical Diameter of Oil Pipeline.   When an oil pipeline is to be installed, a decision is required concerning the diameter of pipe to be used.

The cost of the pipe itself is proportional to the pipe size used. However, the use of pipe having a large diameter reduces the frictional loss, which is reflected in a lower cost for supplying the energy required to maintain the flow of oil.

It is evident that in each of these three cases there is one available size for which the total cost is a minimum; our present problem is to locate this point of correspondence between size and minimum cost by using the available cost data.

In the examples cited above, and in all others of this nature, we shall assume that it is possible to classify all costs pertaining to a particular project into three categories:

1. Costs that are directly proportional to the size selected
2. Costs that are inversely proportional to the size selected
3. Fixed costs, which are independent of the size selected

In reality, costs vary in a manner too complex to permit such rigid classification, but we shall disregard this fact in order to facilitate our investigation of this problem mathematically. In general, let

x = size selected (span length of a bridge, diameter of an oil pipeline, etc.)

y = total cost corresponding to x

C_d = total of all costs directly proportional to x, or $C_d = ax$, where a is a constant

C_i = total of all costs inversely proportional to x, or $C_i = b/x$, where b is a constant

C_f = total of all fixed costs

Then
$$y = C_d + C_i + C_f$$

or
$$y = ax + \frac{b}{x} + C_f$$

We shall now prove that y attains its minimum value when

$$x = \sqrt{\frac{b}{a}} \tag{8-1}$$

and simultaneously
$$C_d = C_i \tag{8-2}$$

Proof. The formula for y can be rewritten as

$$y = ax + bx^{-1} + C_f$$

Differentiating with respect to x,

$$\frac{dy}{dx} = a + (-1)bx^{-2} = a - \frac{b}{x^2}$$

When y is a minimum, this derivative equals zero. Hence, for this particular value,

$$a - \frac{b}{x^2} = 0$$

or

$$x = \sqrt{\frac{b}{a}}$$

We can also write the derivative in the following form:

$$\frac{dy}{dx} = a - \frac{b}{x^2} = \frac{ax}{x} - \frac{b/x}{x} = \frac{C_d}{x} - \frac{C_i}{x}$$

$$\therefore C_d - C_i = x\frac{dy}{dx}$$

Setting this derivative equal to zero, we obtain, when y is a minimum,

$$C_d - C_i = 0$$
or
$$C_d = C_i$$

Equations (8-1) and (8-2), taken in conjunction, are referred to as Kelvin's law. These relationships for minimum cost were derived by Lord Kelvin in connection with the selection of an economical size for an electrical conductor, a problem parallel to that of selecting the economical diameter of an oil pipeline.

In the solution of a numerical problem, we can calculate the value of x corresponding to the minimum value of y either by substituting in Eq. (8-1) for x or by equating the directly varying costs C_d to the inversely varying costs C_i. While both methods produce an identical result, the latter is somewhat simpler to apply, as it does not require the memorization of an equation.

Example 8-1. A steel bridge is to cover a 1,400-ft crossing. The cost of the steel and of an individual pier are estimated to be as follows: $C_s = 600,000 + 7,200S$ and $C_p = 420,000 + 85S$, where C_s is the cost in dollars of the steel, C_p is the average cost in dollars of a pier, and S is the span length. All other costs are independent of the span length. Determine the most economical span length for this bridge.

Solution. The number of piers is one more than the number of spans. The total cost of the piers is as follows:

$$\Sigma C_p = \left(\frac{1,400}{S} + 1\right)(420,000 + 85S)$$

$$= \frac{588,000,000}{S} + 85S + 539,000$$

Let y denote the total cost of steel and piers. By summation,

$$y = 7,285S + \frac{588,000,000}{S} + 1,139,000$$

By Eq. (8-1), y is minimum when

$$S = \sqrt{\frac{b}{a}} = \sqrt{\frac{588,000,000}{7,285}} = 284.1 \text{ ft}$$

$$\text{Number of spans required} = \frac{1,400}{284.1} = 4.93$$

Use 5 spans. The individual span length is 280 ft.

Example 8-2. (P.E. examination problem.) A manufacturer must produce 50,000 parts per year and wishes to select the most economical lot size to be equally spaced during the year. The annual cost of storage and interest on investment varies directly with the lot size and is $1,000 when the entire annual requirement is manufactured in one lot. The cost of setting up and dismantling the machine for each run is $30. What is the most economical lot size?

Solution. Let x = lot size, y = total cost to produce 50,000 units, and $50,000/x$ = number of runs. By proportion, we obtain the following:

$$\text{Annual storage and interest expense} = x\frac{1,000}{50,000} = 0.02x$$

$$\text{Preparation and dismantling cost} = \$30 \text{ per run} = \left(\frac{50,000}{x}\right)30 = \frac{1,500,000}{x}$$

Then

$$y = 0.02x + \frac{1,500,000}{x}$$

When y is a minimum,

$$x = \sqrt{\frac{b}{a}} = \sqrt{\frac{1,500,000}{0.02}} = 8,660 \text{ parts per lot}$$

$$\text{Number of lots required} = \frac{50,000}{8,660} = 5.8$$

If 6 lots are used, lot size is

$$\frac{50,000}{6} = 8,333 \text{ parts}$$

8-3.   Geometrical Derivation of Kelvin's Law.   In Fig. 8-1, the total directly varying cost C_d is represented by the straight line OA, while the total inversely varying cost C_i is represented by the hyperbola BC.

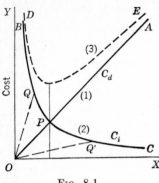

FIG. 8-1

Since the fixed costs have no effect on the location of the point of minimum cost, they will be excluded from our consideration. The summation of C_d and C_i, which we shall term the net cost, is represented by curve DE. As shown in the preceding section, the derivative of C_i with respect to x, which measures the slope of the hyperbola, is

$$\frac{dC_i}{dx} = -\frac{C_i}{x}$$

If a radial line is drawn from the origin to any point Q on the curve, the slope of this line is C_i/x. Hence, at any point,

Slope of radius = −slope of curve

or   Slope of radius = rate of decrease of C_i

When the net cost y attains its minimum value, the slope of the net-cost curve at this minimum point is zero, which means that y is neither increasing nor decreasing at that point. Since y is the summation of the constantly varying elements C_d and C_i, this means that y is a minimum when the rate of increase of C_d (slope of line OA) equals the rate of decrease of C_i (slope of radius). The only point where this condition exists is the intersection point P of the hyperbola and the straight line. Hence, y is a minimum when the total directly varying cost equals the total inversely varying cost, as stated by Kelvin's law.

Parenthetically, it is interesting to observe that the process used above of measuring the slope of a curve by means of the radius to the curve can be applied to any curve having the general equation $y = ax^n$.

Upon differentiating, we obtain

$$\frac{dy}{dx} = nax^{n-1} = na\frac{x^n}{x} = n\frac{y}{x}$$

But the fraction y/x is the slope of the radius at that point. Hence,

$$\text{Slope of curve} = n \times \text{slope of radius}$$

For example, in the cubical parabola shown in Fig. 8-2 ($y = ax^3$), the tangent PT has a slope three times that of the radius OP. It thus follows that the length of the line segment TA is one-third of the abscissa.

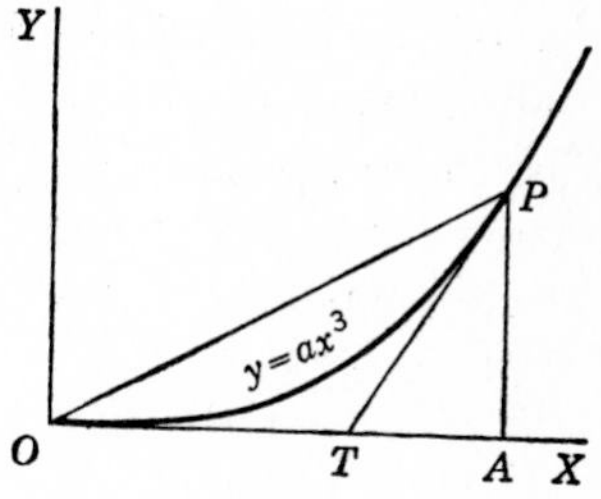

8-4. Straight-line Variation of Costs.
Consider the case of a manufacturing firm producing one standard commodity. Corresponding to each level of production there is a distinct total cost. However, the relationship of this total cost to the number of units produced is a rather complex one, not readily expressible by a mathematical formula.

FIG. 8-2

The total cost consists of myriad elements of cost, each element varying individually with the volume of production. There are certain costs, designated as "overhead," which either remain constant in magnitude or respond only slightly to a variation of production. At the other extreme, there are various costs, such as labor and material, which vary approximately in direct proportion to the volume of production. Between these two extremes are other costs which are fairly constant within a particular range of production but are different for each distinct range.

In many problems, the existence of these intermediate-type costs is ignored for the sake of simplicity. Hence, the total cost of production is considered to consist solely of fixed costs, which are insensitive to the volume of production, and variable costs, which are assumed to vary in direct proportion to production.

In Fig. 8-3, the total cost of production is plotted against the number of units produced in a given period of time, which we shall take as one month. The graph is a straight line. We shall refer to the average cost of the units produced up to a given point as the "unit cost" at that particular point. Consider a straight line to be drawn through the origin, intersecting the cost line at point P. We shall refer to this as a radius. The slope of this line is the quotient obtained from dividing the total cost of production (the ordinate) by the number of units produced (the abscissa). This quotient, however, is the unit cost at point P. We thus arrive at the important property

$$\text{Slope of radius at } P = \text{unit cost at } P$$

Evidently, as production is increased, the unit cost continuously declines. This decline of unit cost, however, is decelerated as production increases.

In problems dealing with the variation of the cost of manufacture, it is generally assumed that all units produced are actually sold and that the unit selling price remains constant. The time value of money is usually disregarded, since we are dealing only with a brief period of time. We shall refer to the average profit realized through the production and sale of a given number of units as the "unit profit" at that stage of production.

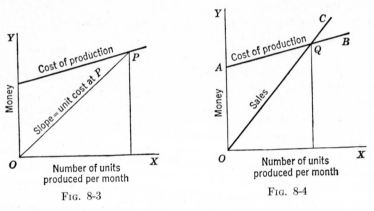

Fig. 8-3

Fig. 8-4

In Fig. 8-4, line OC is drawn to represent the total income obtained by selling the units produced. The slope of this line is the unit selling price. The point of intersection Q of the cost line and sales line is designated as the "break-even point." It denotes the minimum number of units that must be produced in order that income shall exceed cost. Plainly, the greater the fixed costs, the larger the number of units of production required to break even. (The term "break-even point" is also used with other connotations, depending upon the particular problem involved. In general, it is the point of intersection of two graphs.)

Example 8-3. (P.E. examination problem.) Two companies, A and B, manufacture the same article. Company A, relying mostly on machines, has fixed expenses of $12,000 per month and a direct cost of $8 per unit. Company B, using more handwork, has fixed expenses of $4,000 and a direct cost of $20 per unit. At what monthly production rate will the total cost per unit be the same for the two companies? If a unit sells for $30, how many units must each company produce to clear expenses? How do profits compare for 1,000 units per month?

Solution. The variation of the monthly cost of production is represented graphically in Fig. 8-5. Line (1) represents the production cost of company A, while line (2) represents that of company B. Let y denote the cost of production

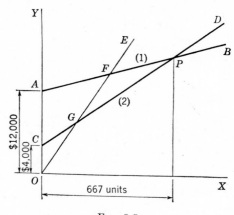

FIG. 8-5

and x the number of units produced. According to the data given, these lines are described by the following equations:

$$\text{Line (1): } y = 12,000 + 8x$$
$$\text{Line (2): } y = 4,000 + 20x$$

Thus, the slope of line (1) is 8, and the slope of line (2) is 20. If a radius OE is drawn, the slope of this line denotes the unit cost at points F and G. Obviously, it is only at the point of intersection P of the two cost lines that the unit cost of the two companies is identical for the same volume of production. To locate P, it is simply necessary to equate the total cost of one company to that of the other. This gives

$$12,000 + 8x = 4,000 + 20x$$
$$12x = 8,000$$
$$x = 667 \text{ units}$$

This result may be regarded as denoting the minimum production required to justify use of the more mechanized system of production. We thus see that there is a range of production (0 to 667 units) for which the nonmechanized system is the more economical of the two; for a volume of production beyond this range, the mechanized system is warranted.

The second part of this problem requires the location of the break-even point for each cost line. Line OH in Fig. 8-6 is the sales line. If S denotes the monthly

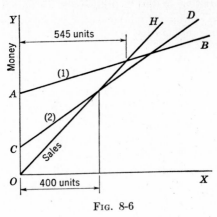

FIG. 8-6

sales, we obtain as the equation of this line

$$S = 30x$$

Equating sales to total cost, we have

Company A

$$S = y$$
$$30x = 12,000 + 8x$$
$$22x = 12,000$$
$$x = 545 \text{ units}$$

Company B

$$S = y$$
$$30x = 4,000 + 20x$$
$$10x = 4,000$$
$$x = 400 \text{ units}$$

Finally, for the third part of the problem, we have the following:

Sales = $30(1,000)$ = \$30,000
Total cost, company A = $12,000 + 8(1,000)$ = \$20,000
Total cost, company B = $4,000 + 20(1,000)$ = \$24,000
Profit, company A = $30,000 - 20,000$ = \$10,000
Profit, company B = $30,000 - 24,000$ = \$6,000

Since the number of units produced (1,000) exceeds the number at which the costs of the two companies are equal (667), it was to be anticipated that the profit of company A would exceed that of company B. In fact, the difference in profit can be computed by multiplying the excess number of units (333) by the difference in slope of the two cost lines, which is 12.

8-5. Linear Programming. Consider that a business firm can manufacture several standard commodities. The production of each commodity consumes a portion of the firm's capital and it makes demands on the production facilities, labor force, and storage space of the firm. Thus, the various commodities are competing with one another for the limited resources of the firm, and therefore the firm must apportion its resources among these commodities in the most efficient manner.

Let A, B, C, . . . denote the commodities that the firm can manufacture, and let N_A, N_B, N_C, . . . denote the number of units of the respective commodity to be produced in a given period of time. The set of values N_A, N_B, N_C, . . . is called the "production schedule" for that period, and a production schedule whose requirements do not exceed the resources of the firm is described as "feasible." It is apparent that there is a group of feasible production schedules and that there is one particular schedule within that group for which the profit is maximum. Consequently, it becomes necessary to devise a technique that will enable us to plot the group of feasible schedules and to identify the schedule of maximum profit.

If production time and cost of production are assumed to vary linearly with the volume of production, the technique under consideration is known as "linear programming." We shall illustrate the simplest form of linear programming in which there are only two competing commodities.

Example 8-4. A firm manufactures two commodities, A and B. The unit cost of production, exclusive of fixed costs, is $10 for A and $7 for B. The unit selling price·is $16 for A and $13.50 for B. The firm is under contract to produce 4,000 units of A and 2,000 units of B each month. In addition to sales covered by this contract, the firm estimates that it can sell a maximum of 5,000 units of each commodity per month. It is the policy of the firm to produce only as many units as can readily be sold.

If production is restricted to one commodity, the factory can turn out 13,000 units of A or 8,500 units of B per month. The capital allotted to monthly production after payment of fixed costs is $100,000.

What monthly production of each commodity will yield the maximum profit?

Solution. The firm operates under the following constraints: the legal contract, potential sales, available capital, and plant capacity. We shall express each constraint mathematically. Let N_A and N_B denote the number of units of A and B, respectively, produced per month. The constraints are as follows:

Legal contract:

$$N_A \geq 4,000 \quad (a) \qquad N_B \geq 2,000 \qquad (b)$$

Potential sales:

$$N_A \leq 9,000 \quad (c) \qquad N_B \leq 7,000 \qquad (d)$$

Available capital:

$$10N_A + 7N_B \leq 100,000 \tag{e}$$

Plant capacity:

$$\text{Time required to produce 1 unit of A} = \frac{1}{13,000} \text{ months}$$

$$\text{Time required to produce 1 unit of B} = \frac{1}{8,500} \text{ months}$$

Then
$$\frac{N_A}{13,000} + \frac{N_B}{8,500} \leq 1$$

or
$$8.5N_A + 13N_B \leq 110,500 \tag{f}$$

In Fig. 8-7, values of N_A and N_B will be plotted along the horizontal and vertical axes, respectively. Viewing the foregoing relationships as equations, we plot the straight lines that represent them, assigning to each line the same identifying letter as its equation. Now considering the sense of each inequality, we find that a set of values of N_A and N_B represented by any point in the shaded area or on its boundaries satisfies the imposed constraints. The shaded area, therefore, is termed the "feasible region." The group of feasible production schedules has now been found, and the problem that remains is to identify the schedule of maximum profit.

Let P denote the monthly profit as computed without allowance for the fixed costs. Comparing the unit cost and selling price of each commodity, we obtain

$$(16 - 10)N_A + (13.50 - 7)N_B = P$$
or
$$6N_A + 6.5N_B = P \tag{g}$$

Arbitrarily setting $P = \$39,000$, we obtain line g in Fig. 8-7. This is called an "equal-profit line" because all points on the line represent production schedules that yield an identical profit (in this case, \$39,000).

Now consider that we assign successively higher values to P. In accordance with Art. 1-7, the equal-profit line is displaced so that it remains parallel to line g and moves outward from the origin. The maximum profit is attained when the equal-profit line is on the verge of leaving the feasible region completely, and it is

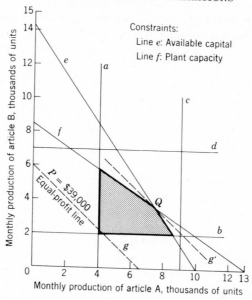

Fig. 8-7

found by drawing lines parallel to line g that Q is the point of maximum profit. The corresponding values of N_A and N_B can be found by scaling or by calculation, in this manner: since Q lies at the intersection of lines e and f, take the corresponding equations and solve them simultaneously. Then

$$10N_A + 7N_B = 100,000$$
and
$$8.5N_A + 13N_B = 110,500$$
Solving,

$$N_A = 7,468 \text{ units}$$
$$N_B = 3,617 \text{ units}$$

8-6. Law of Diminishing Returns. Incremental Costs.

In the preceding material, we have treated problems in which the total cost of production can be resolved into two distinct components, the first component being fixed in amount and the second component varying directly with the volume of production. This condition, however, is an oversimplification of the true state of affairs, for the cost pattern is an intricate one. Each element of cost responds in a unique manner to an increase of production. Moreover, the degree of sensitivity of each element of cost to an increase of production is dependent upon the volume of production that has occurred up to that point. One of the principal factors contributing to this condition is the phenomenon designated as the "law of diminishing returns." This rule states, briefly, that as additional units of labor and capital are applied to a given enterprise, a point is eventually reached beyond which the returns obtained from the additional investment become progressively smaller.

In our present study, we wish to investigate the true variation of cost and profit in relation to the volume of production. We shall assume for this purpose that sufficient data are available to permit the graphical representation of cost by means of a smooth curve.

In the initial stages of production, the unit cost is very high, for the fixed costs are distributed to relatively few units. As production expands, however, the unit cost declines as the fixed costs are absorbed by a constantly growing number of units. At a particular stage of production, the law of diminishing returns commences to operate. Its effect is to produce a continuous increase of unit cost by steadily augmenting the unit variable cost. Thus, the decline of unit cost that occurred in the early stages of production is retarded and eventually arrested entirely at a certain point. Beyond that point, the unit cost rises continuously.

In order to analyze the variation of unit cost and unit profit with regard to the volume of production without encumbering the discussion unduly with precise mathematical terminology, economists find it convenient to study the effect produced at any point by an extension of production by one additional unit. The cost of producing this unit is termed the "incremental cost," "marginal cost," or "incremental investment"; the profit earned by the production and sale of this one unit is termed the "incremental profit" (selling price minus incremental cost), and the rate of return corresponding to this unit is termed the "incremental investment rate" (incremental profit divided by incremental cost). If the number of units produced is very large, the incremental cost corresponding to a particular level of production is approximately equal to the rate of increase of cost at that point, the incremental profit is approximately equal to the rate of increase of profit, and the incremental investment rate is approximately equal to the ratio of the rate of increase of profit to the rate of increase of cost.

In Fig. 8-8a, the cost of production for a given period of time (which we shall take as one month) is plotted against the volume of production for this period, which is measured by the number of units produced. If the radius OC is drawn to a point C on this cost curve, the slope of this radius equals the unit cost at that point. Since we are assuming that each unit is sold directly after it is produced, the straight line OD, whose slope equals the unit selling price, represents the total income obtained from the sale of the units produced. The difference between the ordinates of the sales line and the cost curve represents the profit resulting from the monthly production, as shown in Fig. 8-8b.

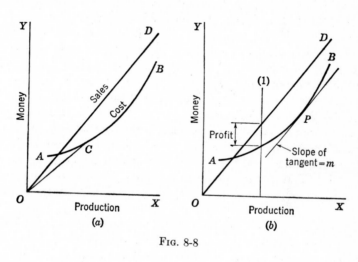

Fig. 8-8

There are three significant points on the cost curve with which we are primarily concerned:

1. The point of maximum profit
2. The point of maximum investment rate
3. The point of maximum economic production

The location of these three points on the cost curve can be obtained by the rational method, based on simple economic logic, or by recourse to mathematics. Furthermore, a mathematical solution can be achieved either by analyzing the geometry of the cost curve or by applying calculus. We shall use all three methods.

Since the cost curve is too complicated to be reduced to a simple mathematical equation, we shall locate the three significant points by graphical construction.

8-7. Point of Maximum Profit. For a particular level of production, the difference between the ordinates of the sales and cost diagrams in Fig. 8-8b represents the profit (or loss) for the month. To determine the point at which this difference between the ordinates is maximum, we proceed as follows: If we start at a particular vertical line, such as line (1), we see that the cost curve is rising less rapidly than the sales line. Hence, the straight line and the curve are diverging from one another, thereby increasing the distance between them. However, as we proceed to the right, since the sales line rises at a constant rate while the cost curve rises at an ever-increasing rate, a point is eventually reached at which the straight line and curve have the same inclination. This is the point at which the vertical distance between them is maximum, for if we continue moving to the right, the straight line and curve begin to converge. Hence, the point of maximum profit is the point at which the cost of production is increasing at the same rate as the monthly sales. To locate this point graphically, it is simply necessary to draw a line parallel to the sales line and tangent to the cost curve. The point of tangency is the point of maximum profit. As an approximation, we may state that the point of maximum profit is the point at which the incremental cost equals the selling price.

The above analysis can be rephrased in economic terms in the following manner: If we start at a level of production represented by line (1), the incremental cost at this level is less than the selling price. Hence, an extension of production results in an enlargement of profit. However, as production expands, the incremental cost continues to rise, until it eventually equals the selling price. If production is extended beyond this point, the incremental cost exceeds the selling price, thereby producing a loss on the additional production. Hence, the point of maximum profit is achieved when the incremental cost equals the selling price.

Finally, we shall develop this property of the point of maximum profit by applying calculus for its determination. Let

x = number of units produced and sold during the month
m = selling price = slope of sales line
s = monthly sales = mx
y = monthly cost of production
v = monthly profit = $s - y = mx - y$

Then $\quad\dfrac{dy}{dx}$ = first derivative of cost with respect to production

= slope of cost curve at a given point

= incremental cost at that point (approximately)

Since v is a function of x, its maximum value is attained when the first derivative of v with respect to x is zero. We therefore have

$$v = mx - y$$
$$\frac{dv}{dx} = m - \frac{dy}{dx}$$

Setting this derivative equal to zero, we obtain

$$\frac{dy}{dx} = m$$

or $\qquad\qquad$ Incremental cost = selling price

Since at this point of maximum profit the incremental cost equals the selling price, the incremental profit is zero.

8-8. Point of Maximum Investment Rate. In Fig. 8-9, a radial line OA is drawn, intersecting the cost curve at points B and C. As we have previously stated, the slope of the radius equals the unit cost at its point of intersection with the cost curve. Thus, the unit cost at points B and C is represented by the slope of line OA. At point B, the slope of the curve is less than the slope of the radius; hence, the incremental cost is less than the unit cost at that point. At point C, on the other hand, the incremental cost exceeds the unit cost.

Now consider the radius to be rotated in a clockwise direction about point O until it is tangent to the cost curve at point P. Since at this point the radius and the tangent line coincide, the unit cost and the incremental cost at point P are equal to one another. Also, P is the point on the curve at which the unit cost (slope of radius) is minimum. We shall now prove that P is also the point of maximum investment rate.

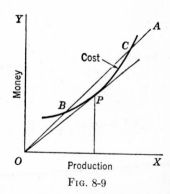

Fig. 8-9

We shall first present this proof on the basis of the rational method. If we start at a particular level of production, such as point B in Fig. 8-9, the cost of producing one additional unit is less than the average cost of the units already produced. Consequently, since the selling price remains constant, the production of this one unit will augment the investment rate. As production is expanded, however, the incremental cost approaches the unit cost until at point P the two are equal. Beyond that point, the incremental cost exceeds the average cost of the units previously produced; consequently, the production of an additional unit will diminish the investment rate. Therefore, the investment rate attains its maximum value when the incremental cost equals the unit cost, which occurs at point P.

Algebraically, we have

$$\text{Investment rate} = \frac{\text{profit}}{\text{investment}}$$
$$= \frac{\text{total sales} - \text{total cost}}{\text{total cost}}$$

Dividing the numerator and denominator of the last fraction by the number of units produced, we have

$$\text{Investment rate} = \frac{\text{selling price} - \text{unit cost}}{\text{unit cost}}$$

or
$$\text{Investment rate} = \frac{\text{selling price}}{\text{unit cost}} - 1$$

The investment rate will therefore be maximum when the unit cost is minimum.

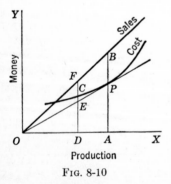

Fig. 8-10

We have thus established the dual properties of the point of maximum investment rate, namely, the equality of unit cost and incremental cost, and the minimal value of its unit cost. It is to be observed that, whereas the point of maximum profit depends for its location upon the selling price, the point of maximum investment rate is independent of this value because it is a function solely of cost.

We shall now apply the geometry of the cost curve to prove that the point of tangency P is the point of maximum investment rate. In Fig. 8-10, the investment rate corresponding to point P is the ratio of the profit

PB to the cost AP. If we select at random any other point on the cost curve, such as point C, then the investment rate at C is the ratio of the profit CF to cost DC. By similar triangles,

$$\frac{PB}{AP} = \frac{EF}{DE}$$

But

$$\frac{EF}{DE} > \frac{CF}{DC}$$

Therefore

$$\frac{PB}{AP} > \frac{CF}{DC}$$

or investment rate at P > investment rate at C. Since this inequality obtains regardless of the location of point C, the investment rate is maximum at P.

Finally, the problem of locating the point of maximum investment rate can be solved by applying calculus. We shall continue the notation previously employed and, in addition, use the letter r to denote the investment rate pertaining to a particular level of production. We then have

$$r = \frac{\text{profit}}{\text{cost}} = \frac{v}{y} = \frac{mx - y}{y} = \frac{mx}{y} - 1$$

Since r is a function of x, it attains its maximum value when the first derivative of r with respect to x equals zero. Therefore, at the point of maximum investment rate,

$$\frac{dr}{dx} = \frac{d}{dx}\left(\frac{mx}{y} - 1\right) = 0$$

$$\frac{my - mx\,(dy/dx)}{y^2} = 0$$

$$\therefore my - mx\frac{dy}{dx} = 0$$

or

$$\frac{dy}{dx} = \frac{y}{x}$$

Hence, at the point of maximum investment rate, the incremental cost equals the unit cost; P is the only point on the cost curve satisfying this condition.

The equation for r derived above can be rewritten in the following manner:

$$r = \frac{m}{y/x} - 1 = \frac{\text{selling price}}{\text{unit cost}} - 1$$

This represents the average investment rate earned by all the units produced up to a particular point.   Let r_i denote the incremental investment rate at this point.   Then

$$r_i = \frac{\text{incremental profit}}{\text{incremental cost}} = \frac{m - dy/dx}{dy/dx}$$

or

$$r_i = \frac{m}{dy/dx} - 1 = \frac{\text{selling price}}{\text{incremental cost}} - 1$$

Since at the point of maximum investment rate the unit cost and incremental cost are equal, the average investment rate and incremental investment rate are also equal.   In summary, therefore, the following properties are possessed by the point of maximum investment rate:

1. Incremental cost equals unit cost.
2. Incremental investment rate equals average investment rate.
3. Unit cost is a minimum.

8-9. Point of Maximum Economic Production.   As long as the production of one additional unit of a commodity yields a profit large enough for the rate of return to exceed the minimum investment rate acceptable to the owners, then the production of this one unit is feasible.   However, as a result of the constant increase of incremental cost that accompanies an expansion of production, a point is eventually reached at which the rate of return resulting from the production of one additional unit exactly equals the minimum acceptable rate.   Hence, production reaches its economical limit at that point at which the incremental investment rate equals this minimum rate required by the investors.

Let q denote the minimum acceptable investment rate.   Setting this equal to the incremental investment rate, we obtain

$$q = r_i = \frac{m}{dy/dx} - 1 \qquad \text{as proved above}$$

Solving for dy/dx,

$$\frac{dy}{dx} = \frac{m}{1 + q}$$

This equation therefore expresses the slope of the cost curve at the point of maximum economic production.

In Fig. 8-11, line (1) is the sales line, having a slope equal to the selling price m; line (2) is a line passing through the origin and having a slope equal to $m/(1 + q)$. P_1 is the point on the cost curve at which the curve is parallel to line (1); it is therefore the point of maximum profit. P_2 is the point on the cost curve at which the curve is parallel to line (2); it is

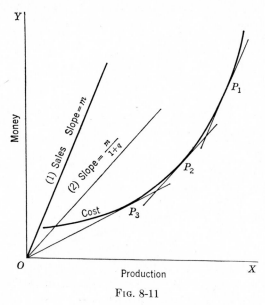

Fig. 8-11

therefore the point of maximum economic production. P_3 is the point on the cost curve at which the radius is tangent to the curve; it is therefore the point of maximum investment rate. We have thus devised a method of locating graphically the three significant points of the cost curve.

8-10. Parabolic Variation of Costs. In order to demonstrate the application of the foregoing principles, we shall assume for illustrative purposes that the cost of production can be expressed by some mathematical formula in terms of the number of units produced during a given period of time. Thus, consider the cost of production y to be related to the number of units produced x by the following parabolic equation:

$$y = ax^2 + bx + c$$

where the coefficients a, b, and c are constants. We then have

$$\text{Incremental cost} = \frac{dy}{dx} = 2ax + b$$

$$\text{Unit cost} = \frac{y}{x} = ax + b + \frac{c}{x}$$

To locate the point of maximum profit, we simply equate the incremental cost to the selling price m, thereby obtaining

$$2ax + b = m$$

or

$$x = \frac{m - b}{2a}$$

To locate the point of maximum investment rate, we equate the incremental cost to unit cost; this gives

$$2ax + b = ax + b + \frac{c}{x}$$

or

$$ax = \frac{c}{x}$$

$$\therefore x = \sqrt{\frac{c}{a}}$$

This result can also be obtained by applying the principle that at the point of maximum investment rate the unit cost is minimum. The derivative of the unit cost with respect to x is

$$a - \frac{c}{x^2}$$

Setting the derivative equal to zero, we obtain the same result as above.

It is interesting to observe that the coefficient b in the formula for y does not influence the location of the point of maximum investment rate, although it does, of course, affect the magnitude of that rate.

Finally, to locate the point of maximum economic production, we have

$$\frac{dy}{dx} = \frac{m}{1 + q}$$

where q represents the minimum acceptable rate of return. Then

$$2ax + b = \frac{m}{1 + q}$$

or

$$x = \frac{m/(1 + q) - b}{2a}$$

Example 8-5. One department of the XYZ Corporation is engaged in the manufacture of a standard article whose selling price is $690. For the normal range of monthly production, the cost of production y is approximately related to the number of units produced x by the following equation:

$$y = 0.5x^2 + 400x + 24,000$$

a. At what volume of production is the maximum profit earned by this department, if we assume that production is actually extended up to this point? What is the investment rate for this volume of production?

b. At what volume of production is this department earning the maximum rate of return on invested capital? What is the rate?

c. If the firm can earn a minimum return of 6 per cent by applying its capital to some other department, what is the maximum number of units that this department should produce per month?

Solution. Part a

$$x = \frac{m - b}{2a} = \frac{690 - 400}{2(0.5)} = 290 \text{ units}$$

Sales = 290 × $690 = $200,100
Cost to produce 290 units = 182,050
Profit = $ 18,050
Investment rate = 9.9 per cent

The accuracy of the first calculation can be verified by calculating the profit corresponding to each volume of production in this region. The results are as follows:

Profit corresponding to 289 units = $18,049.50
to 290 units = 18,050.00
to 291 units = 18,049.50

Solution. Part b

$$x = \sqrt{\frac{c}{a}} = \sqrt{\frac{24,000}{0.5}} = 219 \text{ units}$$

Sales = 219 × $690 = $151,110.00
Cost to produce 219 units = 135,580.50
Profit = $ 15,529.50
Maximum investment rate = 11.5 per cent

Solution.   *Part c*

$$x = \frac{m/(1+q) - b}{2a} = \frac{690/1.06 - 400}{2(0.5)} = 251 \text{ units}$$

This result can be verified by calculating the incremental investment rate at this region, as follows:

Cost to produce 251 units	=	$155,900.50
Cost to produce 250 units	=	155,250.00
Incremental cost	= $	650.50
Incremental profit = $690 − 650.50 = $		39.50
Incremental investment rate = 6.07 per cent		

By a similar calculation, we find the incremental investment rate earned by extending production to 252 units to be 5.91 per cent. Hence, production should be arrested at 251 units. This is true despite the fact that the return obtained on the total investment in producing 251 units is 11 per cent.

PROBLEMS

8-1. (P.E. examination problem.) A company operates two plants, each having a capacity to produce 10,000 units per month. At plant A the fixed costs are $50,000 per month, and the variable costs are $7 per unit. At plant B fixed costs are $70,000 per month, and variable costs are $5 per unit. If 14,000 units are to be produced next month, how should this production be distributed to obtain minimum cost of production?

8-2. (P.E. examination problem.) A package delivery service operates trucks which are capable of giving 60,000 pound-miles service per day. It is found that they are now operating at only half capacity and that the company is now receiving $0.0015 per pound-mile for this service. Daily operating expenses are: overhead, $12; driver, $14; operation, $0.0001 per pound-mile; and repairs and maintenance, $0.90 + $0.00006 per pound-mile. A customer wants a quotation on the pound-mile basis for 12,000 pound miles of delivery service per truck per day. If the package delivery service wants to make an over-all profit of $6 a day per truck, what price should be quoted?

8-3. (P.E. examination problem.) A telephone company finds there is a net profit of $15 per instrument if an exchange has 1,000 or fewer subscribers. If there are over 1,000 subscribers, the profit per instrument decreases 1 cent for each subscriber above that number. How many subscribers would give the maximum net profit?

8-4. (P.E. examination problem.) Two methods are available for recovering ore. One method recovers 75 tons per 100 tons treated at a cost of $3 per ton recovered. The other method recovers 80 tons per 100 tons treated at a cost of $3.25 per ton recovered. If the value of the ore recovered is $6 per ton, which method of recovery should be used? At what value of the recovered ore would it be economical to change the method of recovery?

Ans. Profit realized by treating 100 tons is $225 for first method and $220 for second method. Profits will be equal at a selling price of $7 per ton.

8-5. (P.E. examination problem.) During a slack period a manufacturer can sell 3,000 articles per month, which is two-thirds of the capacity of the factory. The investment in the factory is $200,000, which depreciates at the rate of 5 per cent. Other fixed expenses are $20,000 per year. Maintenance costs vary from 1 per cent at zero output to 4 per cent at full output. If labor and material for the article cost $1 and the article must be sold for $1.20, should the factory stay in production or shut down?

8-6. A manufacturing firm plans to produce annually 3,200 units of a standard article, production to be divided into equal lots spaced at equal intervals. The cost of preparation and dismantling associated with each run is $80. Certain costs are independent of the lot size selected, while the remainder, such as storage and interest, vary in direct proportion to the lot size and amount to $1,280 if all the units are produced in one run. Determine the most economical lot size and the number of runs corresponding thereto.   *Ans.* Four runs of 800 units each

8-7. In the preceding problem, it is apparent that an increase in the preparation and dismantling cost per run will cause an increase in the economical lot size. If it is found that it is most economical to produce the 3,200 units in two lots of 1,600 each, what is the preparation cost?   *Ans.* $320

8-8. A firm manufactures two commodities, A and B. The unit cost of production, exclusive of fixed costs, is $20 for A and $18 for B. The unit selling price is $30 for A and $32 for B. The firm has set the following upper and lower limits to monthly production: for A, 7,000 and 1,000 units; for B, 6,500 and 3,000 units. If production is restricted to one commodity, the plant can produce 5,000 units of A or 7,000 units of B per month. The capital allotted to monthly production after payment of fixed costs is $135,000. To maximize the profit, how many units of each commodity should the firm produce per month?   *Ans.* 1,000 units of A, 5,600 units of B

8-9. A firm manufactures a standard article that sells for $306. The cost of production conforms closely to the following equation:

$$y = 0.3x^2 + 200x + 5,800$$

where x = number of units produced monthly
      y = total monthly cost of production

a. At what volume of production is the investment rate maximum?
                                                       Ans. 139 units
b. At what volume of production is the incremental investment rate 6 per cent?
                                                       Ans. 148 units
c. At what volume of production is the incremental investment rate 5 per cent?
                                                       Ans. 152 units

8-10. A syndicate has available for investment purposes a capital of $200,000, which is to be applied to two investments. The first, designated as investment A, will yield a variable rate of return, depending on the magnitude of the sum invested. The rate of return is estimated to be as follows:

Investment	Annual return	Investment rate (%)
$120,000	$18,400	15.33
130,000	19,500	15.00
140,000	20,400	14.57
150,000	20,950	13.97
160,000	21,300	13.31

Investment B is expected to yield a return of 7 per cent, irrespective of the magnitude of the investment.

Determine the most economical method of apportioning the available capital:

a. By calculating the incremental investment rate corresponding to each increment of investment A.

b. By calculating the total annual return corresponding to every possible division f the capital.   *Ans.* $140,000 for investment A; $60,000 for investment B

CHAPTER 9

THE CONTINUOUS COMPOUNDING OF INTEREST

9-1. The Problem Defined. When a given sum of money is retained in a fund throughout an interest period, then interest is being earned continuously. The interest that accrues, however, remains temporarily dormant rather than dynamic; it does not acquire the self-generating characteristic of money until it is converted to principal at the termination of the interest period. It will therefore be of value to investigate a hypothetical situation in which interest is converted to principal at the very instant it is earned; in other words, where interest is compounded continuously rather than at certain regular intervals.

In Chap. 3, we discussed the method of converting a stipulated nominal interest rate to its effective rate, the latter being defined as the equivalent rate that would apply if interest were compounded only once a year. The relationship between the two rates is expressed by the following equation:

$$r = \left(1 + \frac{j}{m}\right)^m - 1 \qquad (3\text{-}7)$$

where j = nominal interest rate

r = effective interest rate

m = number of compoundings of interest per year

The more frequently interest is compounded, then, of course, the greater is the effective rate. For example, for a nominal rate of 8 per cent per annum, we obtain the following effective rates:

Number of compoundings	Effective rate, %
1	8.000
2	$(1.04)^2 - 1 = 8.160$
4	$(1.02)^4 - 1 = 8.243$
8	$(1.01)^8 - 1 = 8.286$

Although the effective rate increases as the number of compoundings is successively doubled, it is evident that the series of values tabulated above is approaching a definite limit. The value of this limit represents the effective interest rate corresponding to a nominal rate of 8 per cent if interest is compounded continuously. We now wish to determine this value mathematically.

9-2. Significance of the Quantity e. The equation for the effective interest rate with a continuous conversion of interest involves the quantity e, which is perhaps the most important quantity in higher mathematics and science. We shall therefore digress briefly to define and study this quantity.

Consider a variable quantity b to be a function of a quantity a by the relationship

$$b = a^n$$

where a is a positive number. If a has any value greater than 1, and n is made infinitely large, then b increases to infinity. On the other hand, if a has a value less then 1, and n is made infinitely large, b vanishes to zero. Curiosity leads us to inquire: What occurs at the boundary condition, where a equals 1? To answer this question, we shall assign to a a value close to 1 and then consider the variation of b as a is made to approach 1 as a limit. Let

$$b = \left(1 + \frac{1}{n}\right)^n$$

Expanding by the binomial theorem, we obtain

$$\left(1 + \frac{1}{n}\right)^n = 1 + \frac{n}{1!}\frac{1}{n} + \frac{n(n-1)}{2!}\frac{1}{n^2} + \frac{n(n-1)(n-2)}{3!}\frac{1}{n^3} + \cdots$$

$$= 1 + \frac{n}{n}\frac{1}{1!} + \frac{n}{n}\frac{n-1}{n}\frac{1}{2!} + \frac{n}{n}\frac{n-1}{n}\frac{n-2}{n}\frac{1}{3!} + \cdots$$

$$= 1 + \frac{1}{1!} + \left(1 - \frac{1}{n}\right)\frac{1}{2!} + \left(1 - \frac{1}{n}\right)\left(1 - \frac{2}{n}\right)\frac{1}{3!} + \cdots$$

As n increases beyond bound, the sum of this series of terms on the right also increases but approaches a definite limit. This limit is designated as e; that is,

$$e = \lim_{n \to \infty} \left(1 + \frac{1}{n}\right)^n$$

In the above expansion, it is to be noted that as n becomes infinitely large the fractions $1/n$, $2/n$, etc., approach zero as a limit. We therefore have

$$e = \lim_{n \to \infty} \left(1 + \frac{1}{n}\right)^n = 1 + \frac{1}{1!} + \frac{1}{2!} + \frac{1}{3!} + \frac{1}{4!} + \cdots \qquad (9\text{-}1)$$

This is a convergent series, and the value of e, correct to four decimal places, is 2.7183. Consider now that the quantity e is itself raised to the u power, where u is any finite quantity. Applying the definition of e, we have

$$e^u = \lim_{n \to \infty} \left[\left(1 + \frac{1}{n}\right)^n\right]^u = \lim_{n \to \infty} \left(1 + \frac{1}{n}\right)^{un}$$

Expanding this binomial, and again allowing n to become infinitely large, we obtain

$$e^u = 1 + \frac{u}{1!} + \frac{u^2}{2!} + \frac{u^3}{3!} + \frac{u^4}{4!} + \cdots \qquad (9\text{-}2)$$

Finally, consider the binomial $(1 + p/n)$ to be raised to the nth power, where p is any finite quantity. Expanding this binomial gives

$$\left(1 + \frac{p}{n}\right)^n = 1 + \frac{n}{1!}\frac{p}{n} + \frac{n(n-1)}{2!}\left(\frac{p}{n}\right)^2 + \cdots$$

$$= 1 + \frac{p}{1!} + \frac{n}{n}\frac{n-1}{n}\frac{p^2}{2!} + \frac{n}{n}\frac{n-1}{n}\frac{n-2}{n}\frac{p^3}{3!} + \cdots$$

Again allowing n to increase beyond limit, we obtain

$$\lim_{n \to \infty} \left(1 + \frac{p}{n}\right)^n = 1 + \frac{p}{1!} + \frac{p^2}{2!} + \frac{p^3}{3!} + \cdots$$

Comparing this result with Eq. (9-2) above for e^u, it is seen that

$$\lim_{n \to \infty} \left(1 + \frac{p}{n}\right)^n = e^p \qquad (9\text{-}3)$$

9-3. Effective Rate for Continuous Compounding. Let us return now to our problem concerning the continuous compounding of interest. Assume first that the sum of \$1 is earning interest at the nominal rate j, with interest compounded m times per year. Then the principal P_1 at the end of the first year is

$$P_1 = \left(1 + \frac{j}{m}\right)^m$$

As the number of compoundings m becomes infinitely large, then

$$\lim_{m \to \infty} P_1 = \lim_{m \to \infty} \left(1 + \frac{j}{m}\right)^m = e^j$$

Therefore, if interest is compounded continuously,

$$P_1 = e^j$$

and the effective rate r corresponding to the nominal rate j is

$$r = e^j - 1 \qquad (9\text{-}4)$$

For example, for a nominal rate of 8 per cent, we have $r = 8.34$ per cent.

In general, if P_0 is the initial principal and P_n is the principal at the expiration of n years, then, by Eq. (3-2),

$$P_n = P_0(1 + r)^n$$
$$= P_0(1 + e^j - 1)^n$$

or

$$P_n = P_0 e^{jn} \qquad (9\text{-}5)$$

and

$$P_{-n} = P_0 e^{-jn} \qquad (9\text{-}5a)$$

9-4. Analogy of Continuous Compounding with Natural Phenomena.
Equation (9-5) expresses the value at a particular instant of time of a quantity that is expanding at a rate directly proportional to its own magnitude. Although this equation was derived for one particular quantity—a sum of money increasing through the earning of interest as time elapses—it applies to any quantity that expands in this manner, requiring merely a substitution of symbols. To prove that this is so, consider a variable quantity y to be a function of x by the relation

$$y = ce^{kx}$$

where c and k are constants. Substituting kx for u in Eq. (9-2) gives

$$y = c\left(1 + \frac{kx}{1!} + \frac{k^2x^2}{2!} + \frac{k^3x^3}{3!} + \cdots\right)$$

Differentiating both members of the equation with respect to x, we obtain

$$\frac{dy}{dx} = c\left(0 + k + \frac{2k^2x}{2!} + \frac{3k^3x^2}{3!} + \frac{4k^4x^3}{4!} + \cdots\right)$$

$$= kc\left(1 + \frac{kx}{1!} + \frac{k^2x^2}{2!} + \frac{k^3x^3}{3!} + \cdots\right)$$

or $\qquad \dfrac{dy}{dx} = ky$

Thus, as the independent variable x increases, the dependent variable y increases at a rate directly proportional to its own magnitude, with k as the constant of proportionality. Conversely, if it is known that one quantity y varies as a function of x at a rate proportional to its own magnitude, then y is related to x by the equation

$$y = ce^{kx}$$

Let y_0 denote the particular y value when x equals zero. Then

$$y_0 = ce^0 = c$$

We may therefore rewrite the equation for y as

$$y = y_0e^{kx} \qquad (9\text{-}6)$$

If, on the other hand, a variable quantity y decreases with respect to x at a rate directly proportional to its own magnitude and we assign only positive values to the constant k, we have

$$y = y_0e^{-kx} \qquad (9\text{-}7)$$

Figure 9-1 presents the graph of the exponential equation

$$y = ce^{kx}$$

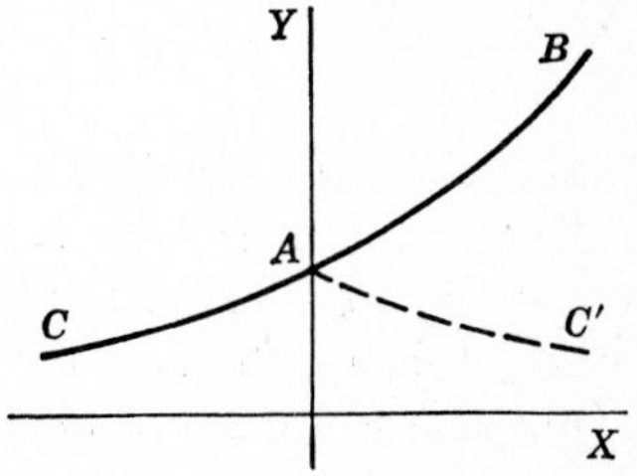

Fig. 9-1

If that portion AC of the curve lying to the left of the YY axis is rotated about that axis through an angle of 180°, it assumes the position AC' in the plane of AB. Thus, if we consider x to have only positive values, AB is the graph of Eq. (9-6) and AC' is the graph of Eq. (9-7).

The relationship

$$y = y_0 e^{kx}$$

is often referred to as "the law of natural growth," for it expresses the mode of increase of a given quantity of matter in which each particle of matter is contributing at a fixed rate to the increase of matter. Thus, the increase of a given principal through the continuous conversion of interest to principal is a perfect illustration. Similarly, Eq. (9-7) can be regarded as the "law of natural decrease," which applies to a situation where each particle of matter is contributing at a fixed rate to the decrease of matter. An illustration of this process is furnished by many types of chemical reaction, in which a substance dissolves at a rate proportional to the quantity present at each instant of time.

There are many processes occurring in natural science and engineering in conformity with these laws. Several of these are enumerated below.

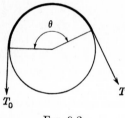

FIG. 9-2

In all cases, the constant of proportionality is represented by the letter k, except as noted. For all cases except the second and the last, the independent variable is time.

Quantity of Bacteria. Under certain prescribed conditions, the bacteria in a culture increase at a rate directly proportional to the number present at any given instant of time. Let N_0 denote the initial number of bacteria in the culture and N_t denote the number present after t time units have elapsed. By Eq. (9-6),

$$N_t = N_0 e^{kt}$$

Belt Friction. In Fig. 9-2, a belt is shown wrapped about a circular cylinder; the angle of contact between the belt and cylinder is θ. A tension of T_0 is applied at one end of the belt and a tension of T at the other, T being the greater of the two. Movement of the belt over the cylinder because of the unequal forces is resisted by friction between the belt and cylinder. Assume, however, that the difference between T and T_0 is of such magnitude that the maximum frictional resistance has been developed, and slipping of the belt impends. Our problem is to determine the value of T in terms of the known quantities T_0, θ, and the coefficient of friction μ.

In textbooks on mechanics, it is proved that as we proceed along the surface of contact the tension in the belt increases at a rate directly proportional to its own magnitude, the constant of proportionality being the frictional coefficient μ. While we shall not present the complete proof here, the basic explanation for this condition is as follows: If we consider a minute section of the belt, the difference in the tensions at the two ends is balanced by the frictional resistance offered to the belt by the cylinder. Consequently, the belt tension is increasing with respect to θ at a rate proportional to the frictional force at that particular point. This force, however, is proportional to the pressure exerted on the cylinder by the belt, which as a result of the curvature of the surface is proportional to the belt tension at that point. The final result of these proportionalities is that the belt tension increases at each point at a rate proportional to itself, with the constant of proportionality equal to μ. In this case, the independent variable is the angle of contact from the initial point to the point under consideration, while the dependent variable is the belt tension at that point. By Eq. (9-6),

$$T = T_0 e^{\mu\theta}$$

Radioactive Disintegration. A radioactive substance disintegrates at a rate proportional to the amount present at any given instant of time. Let R_0 denote the amount existing at some initial date and R_t the amount remaining after t years have elapsed. By Eq. (9-7),

$$R_t = R_0 e^{-kt}$$

This equation was formulated by Rutherford and Soddy in 1903. Although the constant of proportionality k expresses the rate at which the substance disintegrates, physicists prefer to measure this rate by means of the "half-life period." This is the period of time required for the substance to disintegrate to the point at which half the original quantity remains; that is, $R_t = 0.5R_0$. For example, the half-life period of radium is 1,590 years, and that of ionium is 83,000 years.

Electrical Discharge. As a result of leakage, an electrical condenser discharges at a rate proportional to the charge. Hence, at each instant of time, the charge is diminishing at a rate directly proportional to its own magnitude. Let Q_0 denote the initial charge and Q_t denote the charge at the expiration of t units of time. By Eq. (9-7),

$$Q_t = Q_0 e^{-kt}$$

Frictional Resistance. In many forms of motion, the frictional force retarding the motion is directly proportional to the velocity of the moving object at a given instant of time. The deceleration (rate of decrease of motion) varies directly with the force acting; this force in turn varies directly with the velocity. Thus, the velocity continuously decreases in proportion to its own magnitude. Let V_0 denote the initial velocity and V_t denote the velocity at the expiration of t units of time. By Eq. (9-7),

$$V_t = V_0 e^{-kt}$$

Heat Flow. If a body has a temperature higher than that of its surroundings, heat will flow from that body at a rate approximately proportional to the difference in temperature. Hence, the difference in temperature between the body and its surroundings decreases at a rate proportional to itself. Let T_0 denote the initial temperature difference and T_t denote the temperature difference at the expiration of t units of time. By Eq. (9-7),

$$T_t = T_0 e^{-kt}$$

This equation for heat flow is known as Newton's law of cooling.

Atmospheric Pressure. Consider a plane of unit area to be placed in a horizontal position at the earth's surface. The pressure exerted on this plane by the atmosphere equals the weight of the vertical column of air that bears on this plane. If the plane, while remaining parallel to its initial position, is elevated above the earth's surface, the atmospheric pressure on the plane diminishes by an amount equal to the weight of the air through which it has been displaced.

According to the laws of a perfect gas, if the temperature remains constant, the weight of the air contained in a given volume is directly proportional to the pressure of that air. Now, as our plane is elevated a unit height, the drop in pressure on the plane equals the weight of the air contained in the unit volume through which it is displaced; this weight, however, is itself proportional to the atmospheric pressure at that particular elevation. Hence, as the plane is elevated, the atmospheric pressure exerted on the plane decreases at a rate proportional to its own magnitude. Let P_0 denote the atmospheric pressure at the earth's surface and P_z denote the pressure at an elevation of z units. Then, disregarding any variation of temperature, we have by Eq. (9-7),

$$P_z = P_0 e^{-kz}$$

The constant of proportionality k has such value that the atmospheric pressure is halved with each ascent of approximately 3.5 miles.

CHAPTER 10

ANALYSIS OF RANDOM VARIABLES

Statistical analysis based on considerations of probability has assumed an important role in engineering design and industrial engineering, and therefore, it is imperative that the engineer have a working knowledge of this material. This chapter will define and illustrate the basic terms of statistics, and Chap. 11 will explore the subject of probability.

10-1. Definitions. If the value that a given quantity will assume in a given instance cannot be predetermined because it is influenced by chance, this quantity is referred to as a "random" or "stochastic" variable. As an illustration, consider the following situation. A firm manufactures a standard commodity, and studies show that on the average 2.3 per cent of the units are defective. If 100 units are selected at random, how many defective units will the lot contain? The quantity of defective units in the lot is a random variable for the following reasons: the incidence of defective units that is stipulated is merely an average and presumably fluctuates with time, and the number of defective units depends on the particular units that are selected.

If a random variable can assume only discrete (isolated) values, it is called a "discrete" or "step" variable. Thus, in the aforementioned situation, the number of defective units in the lot is a discrete variable with its value restricted to positive integers and zero. On the other hand, if all the values that a random variable can assume form a continuum, the variable is said to be continuous. As an illustration, assume that a ball is thrown against a wall, and define the random variable as the distance from a fixed point on the wall to the point at which the ball strikes the wall. This distance is a continuous random variable.

Many random variables that are continuous in theory must be considered discrete in practice. For example, assume that a firm manufactures rods of a standard length. Owing to manufacturing tolerance, the true length of a rod will presumably differ from the nominal length. The length of a rod selected at random is a random variable, and in theory it is continuous. However, since the precision of measurement is limited, the length of rod must be regarded as discrete. Thus, if the length can be read only to the nearest hundredth of an inch, the step between successive possible values of the length is 0.01 inches. At present, we shall deal solely with discrete variables.

Since the value of a random variable is influenced by chance, the variable acquires a particular value through some act. For example, if the random variable is the length of a rod selected at random, the act consists of selecting a rod and measuring its length. Assume that a set of such acts will be performed, either simultaneously or in succession, thereby generating a set of values of the random variable. It will be assumed that these values are independent of one another insofar as the value assumed by the random variable on one occasion has no bearing on the value it assumes on another occasion.

10-2. Arithmetical Mean. Let x denote a random variable, and assume that n values of x have been generated. These n values may all be distinct, but it is desirable to extract some value that may be considered representative of the entire set. This representative value is called the "arithmetical mean," "mean value," or "average," and it is denoted by x_m. It is defined by the following equation:

$$x_m = \frac{\Sigma x}{n} \tag{10-1}$$

where Σx denotes the algebraic sum of all values of x.

Example 10-1. An examination was given to 6 children, and their scores (in per cent) were as follows: 89, 71, 97, 68, 74, and 87. Find the mean score.

Solution

$$\Sigma x = 89 + 71 + 97 + 68 + 74 + 87 = 486$$

$$x_m = \frac{486}{6} = 81$$

Where the number of x values is very large, the calculation of arithmetical mean can be simplified by assuming a mean and then calculating the necessary correction. Let x_a denote the assumed arithmetical mean.

The relationship between the assumed and the true mean is

$$x_m = x_a + \frac{\Sigma(x - x_a)}{n} \tag{10-2}$$

Example 10-2. Solve Example 10-1 by using an assumed mean grade of 75.

Solution. Refer to the accompanying table.

x	$x - 75$
89	14
71	−4
97	22
68	−7
74	−1
87	12
Total	36

$$x_m = 75 + \frac{36}{6} = 81$$

10-3. Indices of Dispersion. Consider the following sets of values of a random variable:

First set: 18   19   20   21   22
Second set: 4   37   29   2   28

The two sets of values have the common characteristic of having 20 as their mean value, but they differ radically in their basic structure. In the first set, the values cluster within a narrow range; in the second set, the values are widely scattered.

It is frequently necessary to appraise the degree of variability of the values assumed by a random variable. As an illustration, consider again the firm that manufactures rods of a standard nominal length. To control its operations, this firm must determine not simply how close the mean true length lies to the nominal length but also the extent to which the true lengths tend to stray from the mean length. A high degree of variability reflects a low degree of manufacturing precision. Thus, the statistical analyst must devise some means of gaging the dispersion of the values from their mean value.

The arithmetical difference between a given value in a set of values and the mean value of the set is called the "deviation" of that value. Let x_i denote the ith value in the set and d_i its deviation from the mean. By definition,

$$d_i = x_i - x_m \qquad (10\text{-}3)$$

The most direct method of measuring dispersion that suggests itself is to calculate the mean deviation, but this method instantly encounters an insurmountable obstacle: the algebraic sum of the deviations, and therefore the mean deviation, is zero. This can be deduced from Eq. (10-1) or from (10-2) by setting $x_a = x_m$. Therefore, in devising a means of gaging dispersion, use cannot be made of the "raw" deviations.

One method of resolving the problem is to use the absolute values of the deviations. The absolute value of a number is its numerical value without regard to the algebraic sign. The arithmetical mean of the absolute deviations is called the "mean absolute deviation," abbreviated as m.a.d. Let $|d|$ denote the absolute value of a deviation and $|d|_m$ denote the m.a.d. Then

$$|d|_m = \frac{\Sigma|d|}{n} \qquad (10\text{-}4)$$

Example 10-3. Calculate the m.a.d. of the set of test scores in Example 10-1.

Solution. Refer to the accompanying table.

x	$d = x - x_m$	$|d|$
89	8	8
71	-10	10
97	16	16
68	-13	13
74	-7	7
87	6	6
Total	0	60

$$|d|_m = \frac{60}{6} = 10$$

This result signifies that on the average the test scores deviate 10 points from the mean of 81.

The m.a.d. is the most accurate index of dispersion, but the fact that it is composed of absolute values rather than true algebraic values renders it unserviceable in mathematical analysis.

An alternative device that circumvents use of the raw deviations consists of squaring the deviations, thereby obtaining positive values exclusively. The arithmetical mean of the squared deviations is then calculated, and the original process of squaring the deviations is undone by extracting the square root of the mean of the squared deviations. Let σ denote this result. Then

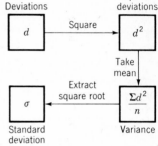

$$\sigma = \sqrt{\frac{\Sigma d^2}{n}} = \sqrt{\frac{\Sigma(x - x_m)^2}{n}} \quad (10\text{-}5)$$

The quantity σ is known as the "standard deviation." The diagram in Fig. 10-1 may be helpful in visualizing the calculation procedure.

Fig. 10-1

Example 10-4. Compute the standard deviation of the set of test scores in Example 10-1.

Solution. Refer to the accompanying table.

x	$d = x - x_m$	d^2
89	8	64
71	-10	100
97	16	256
68	-13	169
74	-7	49
87	6	36
Total		674

$$\sigma = \sqrt{\frac{674}{6}} = \sqrt{112.33} = 10.60$$

Let $(x^2)_m$ denote the arithmetical mean of the squares of the x values. It can be demonstrated that the standard deviation can be calculated by the following alternative equation:

$$\sigma = \sqrt{(x^2)_m - x_m{}^2} \tag{10-5a}$$

Example 10-5. Calculate the standard deviation of the set of test scores in Example 10-1 by applying Eq. (10-5a).

Solution. Refer to the accompanying table.

x	x^2
89	7,921
71	5,041
97	9,409
68	4,624
74	5,476
87	7,569
Total	40,040

$$(x^2)_m = \frac{40,040}{6} = 6,673.33 \qquad x_m{}^2 = 81^2 = 6,561$$

$$\sigma = \sqrt{112.33} = 10.60$$

The advantage of Eq. (10-5a) as compared with Eq. (10-5) is that the former obviates the need for calculating the individual deviations. However, it is advisable to apply both equations as a means of verifying the result.

With reference to Eq. (10-5), the mean of the squared deviations is often of mathematical interest. As shown in Fig. 10-1, this quantity is called the "variance." Let V denote variance. Then

$$V = \frac{\Sigma d^2}{n} = \sigma^2 \tag{10-6}$$

Equations (10-5) and (10-6) are sometimes modified by replacing n in the denominator with $n - 1$. This revision is based on certain theoretical considerations.

PROBLEM

10-1. Given the following sets of values:

First set:	120	106	141	152	123
Second set:	83	104	60	85	77
Third set:	221	245	254	209	189

For each set of values, find the arithmetical mean, the mean absolute deviation, the variance, and the standard deviation.

Ans. For the first set,

$$x_m = 128.40 \qquad |d|_m = 14.48 \qquad V = 263.44 \qquad \sigma = 16.23$$

CHAPTER 11

PROBABILITY

11-1. Introduction. The study of permutations and combinations is a prerequisite to the study of probability and that subject will now occupy our attention.

Relationships pertaining to permutations and combinations can be expressed very compactly by the use of factorial numbers. As defined in Art. 1-5, $n!$ is the product of the integers from 1 to n, inclusive. For our present purpose, it will be more convenient to record the integers in descending order of magnitude. For example, we write

$$4! = 4 \cdot 3 \cdot 2 \cdot 1 = 24$$

The product of a set of consecutive integers can be expressed as the quotient of two factorial numbers. For example,

$$7 \cdot 6 \cdot 5 \cdot 4 = \frac{7 \cdot 6 \cdot 5 \cdot 4 \cdot 3 \cdot 2 \cdot 1}{3 \cdot 2 \cdot 1} = \frac{7!}{3!}$$

For mathematical consistency, it is necessary to assign a value to $0!$ This can be done in the following manner. From the definition of a factorial number, it follows that

$$n! = \frac{(n + 1)!}{n + 1}$$

Setting $n = 0$, we obtain

$$0! = \frac{1!}{1} = \frac{1}{1} = 1$$

We shall apply the following notational system:

$P_{n,r}$ = number of possible permutations of n items taken r at a time

$P_{n,n(j)}$ = number of possible permutations of n items taken all at a time, where j of the n items are alike

$C_{n,r}$ = number of possible combinations of n items taken r at a time

$P(E)$ = probability that an event E will occur

$P(x_i)$ = probability that a random variable x will have the value x_i on a given trial

11-2. Performance of Composite Tasks. A composite task is one that requires the performance of a set of individual tasks. These individual tasks are called the "elements" of the composite task. For example, assume that an individual travels from his home to his office by automobile. The following tasks are some of the components of traveling to work: opening the garage door, driving to the parking lot, and walking from the parking lot to the office.

Example 11-1. A salesman traveling by automobile must drive from town A to town B, and then to town C. He has a choice of 2 roads in going from A to B, and a choice of 4 roads in going from B to C. How many alternative routes does the salesman have in going from A to C?

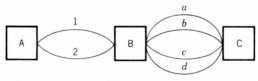

Fig. 11-1

Solution. Refer to Fig. 11-1, where the roads are labeled as shown. There are 8 alternative routes in going from A to C, as follows:

$$
\begin{array}{cccc}
1\text{-}a & 1\text{-}b & 1\text{-}c & 1\text{-}d \\
2\text{-}a & 2\text{-}b & 2\text{-}c & 2\text{-}d
\end{array}
$$

By generalizing from the foregoing result, we arrive at the following statement:

Theorem 11-1. Assume that a composite task has n elements. If the first element can be performed in m_1 alternative ways, the second element in m_2 alternative ways, . . . , the nth element in m_n alternative ways, the composite task can be performed in $m_1m_2m_3 \cdot \cdot \cdot m_n$ alternative ways.

The foregoing theorem can readily be adapted to analogous situations. For example, assume that there are two boxes of crayons. The first box contains red and green crayons, and the second box contains blue, yellow, and brown crayons. If a crayon is drawn at random from each box, the number of alternative color combinations that can be formed is $2 \cdot 3$ or 6.

11-3. Permutations. An arrangement of a group of items in which the order or rank is of significance is called a "permutation." The arrangement can contain the entire group of items or only part of the group.

As an illustration, the following is the complete set of permutations of the first 4 letters of the alphabet taken all at a time.

abcd	bacd	cabd	dabc
abdc	badc	cadb	dacb
acbd	bcad	cbad	dbac
acdb	bcda	cbda	dbca
adbc	bdac	cdab	dcab
adcb	bdca	cdba	dcba

The following is the complete set of permutations of the first 3 letters of the alphabet taken 2 at a time:

ab	ac	bc
ba	ca	cb

Example 11-2. A club must elect a president, secretary, and treasurer. If 3 members have expressed a willingness to serve and each candidate is qualified for all 3 offices, how many alternative slates are available? Assume that each individual is restricted to one office.

Solution. Call the candidates A, B, and C. The possible slates are recorded in the accompanying table. There are 6 alternative slates, each slate constituting a permutation of 3 individuals taken all at a time.

President............	A	A	B	B	C	C
Secretary............	B	C	A	C	A	B
Treasurer............	C	B	C	A	B	A

Assume that we wish to form the complete set of permutations of the first 9 letters of the alphabet taken 4 at a time. How many permutations will the set contain? To answer this question, we reason in this manner.

The first position in the permutation can be assigned to any one of 9 letters; the second position can then be assigned to any one of 8 letters; the third position can then be assigned to any one of 7 letters; etc. Applying Theorem 11-1, we obtain

$$\text{Number of permutations} = 9 \cdot 8 \cdot 7 \cdot 6 = 3{,}024$$

As stated in Art. 11-1, the product of a set of consecutive integers can be expressed as the quotient of two factorial numbers, giving

$$\text{Number of permutations} = 9 \cdot 8 \cdot 7 \cdot 6 = \frac{9!}{5!} = 3{,}024$$

Extending the foregoing reasoning to the general case, we find that

$$P_{n,r} = n(n-1)(n-2) \cdots (n-r+1) = \frac{n!}{(n-r)!} \qquad (11\text{-}1)$$

In the special case where $r = n$, the foregoing equation reduces to

$$P_{n,n} = n! \qquad (11\text{-}1a)$$

Example 11-3. There are 7 children in a room, and there are 4 seats in a row. Children are to be assigned to these seats.

a. How many seating arrangements can be devised?

b. How many seating arrangements can be devised if Charles and Mary cannot be seated alongside one another?

Solution. Part a

$$P_{7,4} = 7 \cdot 6 \cdot 5 \cdot 4 = 840$$

Part b. We shall determine the number of seating arrangements that violate the imposed restriction. For this purpose, assume the reverse of what is stipulated, namely, that Charles and Mary must both be assigned seats and that they must be seated alongside one another.

Now consider that Charles and Mary are coupled to form a unit, and assume tentatively that Charles must precede Mary. The two adjacent seats that these two children will occupy may be said to constitute a single *position*. Thus, the original 4 seats are reduced to 3 positions. The Charles-Mary unit can be assigned to any one of 3 positions. There then remain 2 vacant positions and 5 children. Therefore, the seating arrangement can be completed in $5 \cdot 4$ distinct ways. Thus, the number of arrangements in which Charles immediately precedes Mary is $3 \cdot 5 \cdot 4$ or 60. We now interchange Charles and Mary, thereby obtaining another set of 60 arrangements.

We have found that the number of unacceptable arrangements is 120, and therefore the number of acceptable arrangements is

$$840 - 120 = 720$$

Assume that we wish to form the complete set of permutations of the 8 letters of the word *parabola,* taken all at a time. This word contains 6 distinct letters (a, b, l, o, p, and r), and the letter a occurs 3 times. How many permutations will there be? We reason as follows.

One permutation in the set is

rbaopaal

An interchange of the a's will not yield a new permutation. Assume, however, that we make the a's distinct and call them a_1, a_2, and a_3. The foregoing permutation is then transformed to the following group of permutations:

$$
\begin{array}{ll}
rba_1opa_2a_3l & rba_1opa_3a_2l \\
rba_2opa_1a_3l & rba_2opa_3a_1l \\
rba_3opa_1a_2l & rba_3opa_2a_1l
\end{array}
$$

Thus, if the a's are made distinct, each permutation in the original set is transformed to a group of $P_{3,3}$ new permutations. The new set, with the a's all distinct, contains of course $P_{8,8}$ permutations.

Let $P_{n,n(j)}$ denote the number of permutations of n items taken all at a time, where j of the n items are alike. We have found that

$$P_{8,8(3)} \times P_{3,3} = P_{8,8}$$

$$\therefore P_{8,8(3)} = \frac{P_{8,8}}{P_{3,3}} = \frac{8!}{3!}$$

Alternatively, the number of permutations can be found in this manner: the letter b can be placed in any one of 8 positions, l in any one of 7 positions, o in any one of 6 positions, p in any one of 5 positions, and r in any one of 4 positions. The a's can then be placed only in the remaining positions, and their relative order is immaterial. Thus,

$$P_{8,8(3)} = 8 \cdot 7 \cdot 6 \cdot 5 \cdot 4 = \frac{8!}{3!}$$

In general,
$$P_{n,n(j)} = \frac{n!}{j!} \qquad (11\text{-}2)$$

Now assume that we wish to form the complete set of permutations of the 8 letters of the word *parallel*, taken all at a time. This word contains 5 distinct letters (a, e, l, p, and r); the letter a occurs twice and the letter l occurs 3 times. To find the number of permutations that can be formed, we start with the permutation

$$ealplarl$$

Now we make the a's distinct and call them a_1 and a_2. The foregoing permutation is transformed to the following pair of permutations:

$$ea_1pla_2rl \qquad ea_2pla_1rl$$

Now we make the l's distinct and call them l_1, l_2, and l_3. The first permutation in this pair is transformed to the following:

$$ea_1l_1pl_2a_2rl_3 \qquad ea_1l_2pl_1a_2rl_3 \qquad ea_1l_3pl_1a_2rl_2$$
$$ea_1l_1pl_3a_2rl_2 \qquad ea_1l_2pl_3a_2rl_1 \qquad ea_1l_3pl_2a_2rl_1$$

The final set, with the a's and l's all distinct, contains of course $P_{8,8}$ permutations.

Let $P_{n,n(j,k)}$ denote the number of permutations of n items taken all at a time, where j of the n items are alike and k other items are alike. We have found that

$$P_{8,8(2,3)} \times P_{2,2} \times P_{3,3} = P_{8,8}$$
$$\therefore P_{8,8(2,3)} = \frac{P_{8,8}}{P_{2,2} \times P_{3,3}} = \frac{8!}{2!3!}$$

In general,
$$P_{n,n(j,k)} = \frac{n!}{j!k!} \tag{11-2a}$$

Now consider the special case in which the n items are divided into two groups of like items. Let j denote the number of items in one group and $n - j$ the number of items in the other group. Equation (11-2a) becomes

$$P_{n,n(j,n-j)} = \frac{n!}{j!(n-j)!} \tag{11-2b}$$

Example 11-4. A box contains 5 balls, of which 3 are red and 2 are green. If the 5 balls are to be placed in a row, how many different arrangements by color can be formed? Verify the answer by recording the arrangements, using the letters R and G to denote red and green, respectively.

Solution

$$P_{5,5(3,2)} = \frac{5!}{3!2!} = \frac{5 \cdot 4}{2 \cdot 1} = 10$$

The permutations are generated methodically in the accompanying table. The permutations on line 1 are formed by starting with *R-R-R-G-G* and then successively displacing the first *G* one place to the left. The permutations on lines 2, 3, and 4 are then formed by starting with the permutation on line 1 of each column and then successively displacing the second *G* one place to the left, until the two *G*'s occupy adjoining positions.

1	*R-R-R-G-G*	*R-R-G-R-G*	*R-G-R-R-G*	*G-R-R-R-G*
2		*R-R-G-G-R*	*R-G-R-G-R*	*G-R-R-G-R*
3			*R-G-G-R-R*	*G-R-G-R-R*
4				*G-G-R-R-R*

11-4. Combinations. A grouping of items in which the order or rank is of no significance, or in which the order or rank is predetermined, is called a "combination." For example, assume that we form combinations of the first 6 letters of the alphabet, taken 4 at a time. There are 15 possible combinations as shown in Table 11-1.

TABLE 11-1. COMBINATIONS OF FIRST SIX
LETTERS TAKEN FOUR AT A TIME

1	*abcd*	*bcde*	*cdef*
2	*abce*	*bcdf*	
3	*abcf*	*bcef*	
4	*abde*	*bdef*	
5	*abdf*		
6	*abef*		
7	*acde*		
8	*acdf*		
9	*acef*		
10	*adef*		

Similarly, assume that we form three-digit numbers by taking the digits from 1 to 7, inclusive, and arranging them in descending order of magnitude. The following are some of the numbers that can be formed in the prescribed manner:

$$652 \qquad 731 \qquad 432 \qquad 643$$

Each number thus formed constitutes a combination of the digits 1 through 7 taken 3 at a time.

The basic distinction between a permutation and a combination is as follows. In forming a permutation of n items taken r at a time, we are concerned with both the identity of the r items selected and their relative position or rank. On the other hand, in forming a combination of n items taken r at a time, we are concerned solely with the identity of the r items selected.

There is a simple relationship between the number of possible combinations and the number of possible permutations. To develop this relationship, assume that we have formed the entire set of combinations of the first 5 letters of the alphabet taken 3 at a time. This set of combinations can be transformed to the entire set of permutations by rearranging the letters in each combination in as many ways as possible. For example, the combination *bce* yields the following permutations:

$$bce \qquad bec \qquad cbe \qquad ceb \qquad ebc \qquad ecb$$

Thus, to every combination there corresponds a group of $P_{3,3}$ permutations. Then

$$C_{5,3} \times P_{3,3} = P_{5,3}$$

$$\therefore C_{5,3} = \frac{P_{5,3}}{P_{3,3}} = \frac{5!/(5-3)!}{3!} = \frac{5!}{3!(5-3)!}$$

$$= \frac{5!}{3!2!} = \frac{5 \cdot 4}{2 \cdot 1} = 10$$

In general, $$C_{n,r} = \frac{n!}{r!(n-r)!} \qquad (11\text{-}3)$$

Example 11-5. An instructor has prepared a set of 12 problems and will include 8 of these problems in an examination. How many different examinations can be formed?

Solution

$$C_{12,8} = \frac{12!}{8!4!} = \frac{12 \cdot 11 \cdot 10 \cdot 9}{4 \cdot 3 \cdot 2 \cdot 1} = 495$$

Example 11-6. A club has 18 members and a committee consisting of 5 members is to be formed. If all members are qualified to serve on the committee and the members of the committee will be of equal rank, in how many ways can the committee be formed?

Solution. Since the committee members are of equal rank, we are concerned solely with the identity of these members. Therefore, each committee constitutes a combination of 18 individuals taken 5 at a time.

$$C_{18,5} = \frac{18!}{5!\,13!} = \frac{18 \cdot 17 \cdot 16 \cdot 15 \cdot 14}{5 \cdot 4 \cdot 3 \cdot 2 \cdot 1} = 8{,}568$$

With respect to Eq. (11-3), if we set $r = 0$ we obtain

$$C_{n,0} = 1$$

The combination having 0 items may be regarded as the "null" combination, and there is only one null combination.

Three corollaries stem from Eq. (11-3). First, since the quantities r and $n - r$ can be interchanged in this equation without changing the result, it follows that

$$C_{n,n-r} = C_{n,r} \tag{11-4}$$

This equality was to be expected, for whenever r of the n items are selected for inclusion in the combination, the $n - r$ excluded items form a corresponding combination. Thus, to every combination having r items there corresponds a combination having $n - r$ items and vice versa.

Secondly, since Eq. (11-3) is analogous to Eq. (11-2*b*), it follows that

$$P_{n,n(j,\,n-j)} = C_{n,j} \tag{11-5}$$

This equality was also to be expected. Consider that the n items to be permuted consist of j items of one class and $n - j$ items of another class. In selecting positions in the permutation for the j items in the first class, we are forming a combination of n positions taken j at a time. The remaining positions can only be filled by the items of the second class.

Finally, a comparison of Eq. (11-3) with Eq. (1-10) reveals that $C_{n,r}$ is the coefficient in the $(r + 1)$st term of the binomial expansion. For example, the coefficient in the fourth term of the binomial expansion is

$$\frac{n(n - 1)(n - 2)}{3!} = \frac{n!}{3!(n - 3)!} = C_{n,3}$$

Therefore, by setting $a = b = 1$ in Eq. (1-10), we obtain

$$C_{n,0} + C_{n,1} + C_{n,2} + C_{n,3} + \cdots + C_{n,n} = 2^n \qquad (11\text{-}6)$$

11-5. Determination of $C_{n,r}$ by Pascal's Triangle.   There is a simple recursive formula that enables us to generate values of $C_{n,r}$ by a chain process.   Refer to Table 11-1, which presents all combinations of the first 6 letters of the alphabet taken 4 at a time.   The first column was constructed by combining the first letter with 3 of the subsequent 5 letters, the number of combinations being $C_{5,3}$.   The second column was constructed by combining the second letter with 3 of the subsequent 4 letters, the number of combinations being $C_{4,3}$.   The third column was constructed by combining the third letter with the 3 subsequent letters, the number of combinations being $C_{3,3}$.   Therefore,

$$C_{6,4} = C_{5,3} + C_{4,3} + C_{3,3} \qquad (a)$$

and in general,

$$C_{n,r} = C_{n-1,r-1} + C_{n-2,r-1} + \cdots + C_{r-1,r-1} \qquad (11\text{-}7)$$

Setting $n = 5$ and $r = 4$, we obtain

$$C_{5,4} = C_{4,3} + C_{3,3} \qquad (b)$$

Combining Eqs. (a) and (b),

$$C_{6,4} = C_{5,3} + C_{5,4}$$

and in general,

$$C_{n,r} = C_{n-1,r-1} + C_{n-1,r} \qquad (11\text{-}8)$$

This relationship is found to be consistent with Eq. (11-3).

Equation (11-8) affords a means of generating values of $C_{n,r}$, using the format shown in Table 11-2.   Values of n are read in the column at the left; values of r are read in the row at the top.   To illustrate the procedure, assume that the table has been constructed up to and including the horizontal row corresponding to $n = 7$.   According to the data in the table,

$$C_{7,0} = 1 \qquad C_{7,1} = 7 \qquad C_{7,2} = 21 \qquad C_{7,3} = 35$$

and so on. We now set $n = 8$ and apply Eq. (11-8) repeatedly to obtain the following values:

$$C_{8,0} = C_{7,0} = 1$$
$$C_{8,1} = C_{7,0} + C_{7,1} = 1 + 7 = 8$$
$$C_{8,2} = C_{7,1} + C_{7,2} = 7 + 21 = 28$$
$$C_{8,3} = 21 + 35 = 56 \qquad C_{8,4} = 35 + 35 = 70$$

and so on. Table 11-2 can be continued indefinitely, the only limitation being the available space. The triangular array of numbers in this table is known as "Pascal's triangle."

TABLE 11-2. VALUES OF $C_{n,r}$ AND $P_{n,n(r,n-r)}$

n	r									
	0	1	2	3	4	5	6	7	8	9
0	1									
1	1	1								
2	1	2	1							
3	1	3	3	1						
4	1	4	6	4	1					
5	1	5	10	10	5	1				
6	1	6	15	20	15	6	1			
7	1	7	21	35	35	21	7	1		
8	1	8	28	56	70	56	28	8	1	
9	1	9	36	84	126	126	84	36	9	1

11-6. Definition of Probability. A "trial" is an act or process whose outcome is influenced by chance. To illustrate a trial, consider the standard type of cubical die, which has dots on each of its six faces. One face has 1 dot, another has 2 dots, another has 3 dots, etc. When the die is tossed, the manner in which it lands is determined by chance. We define the "outcome" as the number of dots on the face that lands on top. Therefore, there are 6 possible outcomes, namely, the integers from 1 to 6, inclusive.

Assume that a trial has n possible outcomes, which we designate as $O_1, O_2, O_3, \ldots, O_n$. If one outcome is just as likely to occur as any other, we define the "probability" of a particular outcome O_i as $1/n$. Symbolically, this is written as

$$P(O_i) = \frac{1}{n} \qquad (11\text{-}9)$$

An "event" is a specified outcome or set of outcomes. For example, with respect to tossing a die, we may define the following events:

Event E_1: The outcome is 4.

Event E_2: The outcome is even. This event comprises the outcomes 2, 4, and 6.

Event E_3: The outcome is at least 3. This event comprises the outcomes 3, 4, 5, and 6.

Two events are said to be "mutually exclusive" if either one, but not both, can result from a single trial. For example, consider the following pair of events:

Event E_1: The outcome is even.

Event E_2: The outcome is odd.

These events are mutually exclusive. Now consider the following pair:

Event E_3: The outcome is even.

Event E_4: The outcome is less than 4.

There is one outcome (namely, 2) that satisfies both events. Therefore, both events can result from a single trial, and they are not mutually exclusive.

11-7. Laws of Probability. The calculation of probability requires an application of several simple laws, which we shall now present.

Theorem 11-2. Assume that a trial has n possible outcomes of equal probability. If any one of r outcomes will produce an event E, the probability that E will occur is

$$P(E) = \frac{r}{n}$$

For example, with respect to tossing a die, consider this event: the outcome is odd and more than 2. This event is satisfied by 2 of the 6 possible outcomes (namely, 3 and 5). Therefore, the probability that this event will occur is 2/6 or 1/3.

Now consider the following events with respect to tossing a die:

Event E_1: The outcome is 1, 2, 3, 4, 5, or 6. In accordance with Theorem 11-2, the probability of this event is 6/6 or 1. However, since this event includes all possible outcomes, the event is certain to occur. Therefore, an event that is certain has a probability of 1.

Event E_2: The outcome is 7. Since this event is not satisfied by any of the possible outcomes, its probability is 0. Moreover, the event cannot possibly occur. Therefore, an impossible event has a probability of 0.

In summary, we can say that the probability of an event can range from 0 to 1. The lower bound corresponds to an impossible event, and the upper bound corresponds to an event that is certain. The probability of a specified event can be expressed in the form of an ordinary fraction, a decimal fraction, or a per cent, whichever is more suitable in a given situation.

Theorem 11-3.   If two events E_1 and E_2 are mutually exclusive, the probability that *either E_1 or E_2* will occur is the sum of their respective probabilities.

Proof.   Assume the following: there are n possible outcomes of equal probability. There is a set of r_1 outcomes such that any one of them will produce E_1. There is another set of r_2 outcomes, wholly distinct from the first, such that any one of them will produce E_2. Then

$$P(E_1 \text{ or } E_2) = \frac{r_1 + r_2}{n} = \frac{r_1}{n} + \frac{r_2}{n} = P(E_1) + P(E_2)$$

In the subsequent material, it is to be understood that if there is a group of items and one item is to be selected at random, all items in the group have equal likelihood of being selected.

Example 11-7.   A box contains 25 chips. Of these, 9 bear the number 1, 6 bear the number 2, 8 bear the number 3, and 2 bear the number 4. If a chip is drawn at random, what is the probability that it bears an odd number?

Solution.   Although the answer is apparent, we shall nevertheless determine it by a formal application of Theorem 11-3. Let E_1 and E_2 denote, respectively, drawing a chip marked 1 and drawing a chip marked 3. Then

$$P(E_1) = \frac{9}{25} \qquad P(E_2) = \frac{8}{25}$$

The specified event (drawing a chip that bears an odd number) occurs if either E_1 or E_2 occurs. Then

$$P(\text{odd}) = \frac{9}{25} + \frac{8}{25} = \frac{17}{25} = 0.68$$

Consider that there are two boxes containing 12 balls each. Of these, 3 are blue, 4 are red, and 5 are green. A ball will be drawn at random from the first box and discarded. Then a ball will be drawn at random from the second box. We shall refer to the drawing of a ball from the first box as the first trial, and the drawing of a ball from the second box as the second trial. What is the probability of drawing a green ball on the second trial? The answer of course is $5/12$.

Now consider that the ball drawn from the first box, instead of being discarded, is placed in the second box before a ball is drawn from that box. Now what is the probability of drawing a green ball on the second trial? This question cannot be answered unequivocally, for there are two possible answers. When a ball has been drawn from the first box and placed in the second box, there will be 13 balls in the second box. If the ball drawn from the first box is blue or red, there will still be 5 green balls in the second box, and the probability of drawing a green ball will be $5/13$. On the other hand, if the ball drawn from the first box is green, there will then be 6 green balls in the second box, and the probability of drawing a green ball will be $6/13$. Therefore, this problem is ambiguous.

In the first situation, where the ball drawn from the first box was discarded, the second trial was completely independent of the first. The probability of a particular outcome in the second trial was not influenced by the outcome of the first trial. In the second situation, however, the second trial was dependent on the first. The probability of a particular outcome in the second trial cannot be determined until the outcome of the first trial is known. (This is referred to as a "conditional probability.") Therefore, in all instances in which there will be multiple trials, it is necessary to consider whether the trials are independent or dependent.

Theorem 11-4. Assume that a trial T_1 has n_1 possible outcomes of equal probability and that a trial T_2, independent of T_1, has n_2 possible outcomes of equal probability. The trials may be performed in sequence or simultaneously.

a. The number of possible combined outcomes of the two trials is $n_1 n_2$, all of equal probability.

b. If an event E_1 can be produced by any one of r_1 outcomes of T_1 and an event E_2 can be produced by any one of r_2 outcomes of T_2, then *both E_1 and E_2* can be produced by any one of $r_1 r_2$ combined outcomes.

Proof. Both statements follow from Theorem 11-1. For example, if there are r_1 ways of producing E_1 and r_2 ways of producing E_2, then there are $r_1 r_2$ ways of producing both E_1 and E_2.

Theorem 11-5. Assume that there will be two independent trials. The probability that the first trial will produce an event E_1 and the second trial will produce an event E_2 is the product of their respective probabilities.

Proof. Applying the notation of Theorem 11-4, we have

$$P(E_1) = \frac{r_1}{n_1} \qquad P(E_2) = \frac{r_2}{n_2}$$

There are $n_1 n_2$ possible combined outcomes of equal probability, and $r_1 r_2$ of these meet the requirement. Then

$$P(E_1 \text{ and } E_2) = \frac{r_1 r_2}{n_1 n_2} = \frac{r_1}{n_1} \frac{r_2}{n_2} = P(E_1) \times P(E_2)$$

Example 11-8. A die is to be tossed twice. What is the probability of obtaining a number less than 3 on the first toss and an odd number on the second toss? Verify the answer by recording the satisfactory combined outcomes.

Solution. The first event is satisfied by the outcomes 1 and 2, and the second event is satisfied by the outcomes 1, 3, and 5. Then

$$P(E_1) = \frac{2}{6} \qquad P(E_2) = \frac{3}{6}$$

$$P(E_1 \text{ and } E_2) = \frac{2}{6} \frac{3}{6} = \frac{1}{6}$$

Check. The combined outcomes that cause both E_1 and E_2 to occur are as follows:

$$
\begin{array}{ccc}
1\text{-}1 & 1\text{-}3 & 1\text{-}5 \\
2\text{-}1 & 2\text{-}3 & 2\text{-}5
\end{array}
$$

Thus, there are $2 \cdot 3$ or 6 satisfactory combined outcomes. The number of possible combined outcomes is $6 \cdot 6$ or 36. Then

$$P(E_1 \text{ and } E_2) = \frac{6}{36} = \frac{1}{6}$$

Theorem 11-6. Assume that there will be two trials, the second dependent on the first. The probability that the first trial will yield an event E_1 and the second trial will yield an event E_2 is the product of their respective probabilities, where the probability of E_2 is calculated on the premise that E_1 has occurred.

The validity of this statement can most effectively be demonstrated by analyzing a particular situation. Consider that a box contains 9 balls, 5 of which are green. A ball will be drawn at random and discarded. Then a second ball will be drawn at random. What is the probability that both balls will be green? To produce the specified event, it is possible to draw any one of 5 balls on the first trial and any one of 4 balls on the second trial. Thus, there are $5 \cdot 4$ ways of drawing the two green balls. Without imposing any restriction concerning color, it is possible to draw any one of 9 balls on the first trial and any one of 8 balls on the second trial. Thus, there are $9 \cdot 8$ ways of drawing two balls. The probability that both balls will be green is

$$P(\text{both green}) = \frac{5 \cdot 4}{9 \cdot 8} = \frac{5}{9}\frac{4}{8} = \frac{5}{18}$$

Now, the probability of drawing a green ball on the first trial is 5/9. If this event has in fact occurred, the probability of then drawing a green ball on the second trial is 4/8. Therefore, the probability that both balls are green equals the product of the following: the probability that the first ball is green and the probability that the second ball is green if the first ball drawn was green. Applying this reasoning to the general case, we arrive at Theorem 11-6.

Parenthetically, it should be observed that the foregoing situation can be analyzed by an alternative approach. Since there are 9 balls in the box and 2 will be selected, we are forming combinations of 9 items taken 2 at a time; the number of possible combinations is $C_{9,2}$. Since there are 5 green balls, the number of satisfactory combinations is $C_{5,2}$. Referring to Table 11-2, we obtain

$$P(\text{both green}) = \frac{C_{5,2}}{C_{9,2}} = \frac{10}{36} = \frac{5}{18}$$

Example 11-9. A box contains 5 blue and 4 green balls and another box contains 11 blue and 8 green balls. A ball will be drawn at random from the first box and placed in the second box. Then a ball will be drawn from the second box.

a. What is the probability that a blue ball will be drawn from both boxes?

b. What is the probability that a green ball will be drawn from the first box and a blue ball from the second box?

Solution. Initially, there are 9 balls in the first box and 19 balls in the second box.

Part a

$$P(\text{blue,blue}) = \frac{5}{9}\frac{12}{20} = \frac{1}{3}$$

Part b

$$P(\text{green,blue}) = \frac{4}{9}\frac{11}{20} = \frac{11}{45}$$

Example 11-10. A box contains 9 balls; 4 are green, 3 are blue, and 2 are yellow. A ball will be drawn at random from the box and then returned to the box, this process being performed 5 times. What is the probability (to three decimal places) that a green ball will be drawn exactly twice?

Solution. The trials are independent of one another. By combining events, we establish a hierarchy of events, in this manner:

First-order event: A green ball is drawn on a single trial.

Second-order event: A green ball is drawn exactly twice in 5 trials, in some definite sequence.

Third-order event: A green ball is drawn exactly twice in 5 trials, in any sequence whatever.

Let the letters G and N refer to "green" and "not green," respectively. For the first-order event,

$$P(G) = \frac{4}{9} \qquad P(N) = \frac{5}{9}$$

Consider the following pair of second-order events:

Trial:	1	2	3	4	5
Event E_1..........	G	G	N	N	N
Event E_2..........	N	N	G	N	G

By Theorem 11-5,

$$P(E_1) = \frac{4}{9}\frac{4}{9}\frac{5}{9}\frac{5}{9}\frac{5}{9} = \left(\frac{4}{9}\right)^2\left(\frac{5}{9}\right)^3$$

$$P(E_2) = \frac{5}{9}\frac{5}{9}\frac{4}{9}\frac{5}{9}\frac{4}{9} = \left(\frac{4}{9}\right)^2\left(\frac{5}{9}\right)^3$$

Thus, all second-order events have equal probability.

How many second-order events are possible? In the foregoing table, there are 5 positions; 2 are occupied by G and 3 are occupied by N. It follows that the number of second-order events is $P_{5,5(2,3)}$. By Eq. (11-2b), this number is 5!/2!3! The second-order events are mutually exclusive, and therefore, the probability of the third-order event is the sum of the probabilities of the second-order events. Then

$$P(G\text{-}G) = \frac{5!}{2!3!}\left(\frac{4}{9}\right)^2\left(\frac{5}{9}\right)^3 = 10\left(\frac{4^2 \cdot 5^3}{9^5}\right) = 0.339$$

By applying the method of analysis used in Example 11-10 to the general case, we arrive at the following.

Theorem 11-7. If p is the probability that an event will occur in a single trial, the probability $P(r)$ that the event will occur exactly r times in n independent trials is

$$P(r) = \frac{n!}{r!(n-r)!}\, p^r(1-p)^{n-r}$$

Reference to Eq. (1-10) reveals that this expression is the $(r+1)$st term in the binomial expansion

$$[(1-p)+p]^n$$

Therefore,

$$P(0)+P(1)+P(2)+\cdots+P(n) = [(1-p)+p]^n = 1^n = 1$$

as was to be expected.

11-8. Empirical Determination of Probability. Assume that a box contains 7 green and 3 red balls. If a ball is drawn at random, the probability that this ball is green is of course 0.7.

Now consider that the ball that is drawn is returned to the box and that this process is repeated indefinitely. Consider also that we count the number of drawings and the number of times a green ball is drawn, and compute the ratio of the second number to the first. This ratio will assume a new value with each drawing. However, having confidence in our mathematical formulation of probability, we assume that as the number of drawings increases beyond bound this ratio approaches 0.7 as a limit.

Reversing our point of view, assume that it is known simply that a box contains 10 balls and that each ball is either green or red. If we find, as the number of drawings becomes extremely large, that the ratio of the number of green balls drawn to the number of drawings is apparently approaching 0.7 as its limit, we may safely conclude that the box contains 7 green balls.

Thus far, we have confined our discussion to trials for which the probability of a specified event can be calculated precisely. However, there are many types of trials that do not lend themselves to mathematical analysis. The random variations that govern probability may be too numerous, too complex, or simply unknown. For example, what is the probability that a ten-year-old American child selected at random will secure a grade of 75 per cent in a newly devised examination? If this examination is not comparable to any previously given, no satisfactory estimate of probability can be offered. However, if this examination is administered to a large sample of carefully selected ten-year-old American children and their grades are compiled and averaged, it becomes possible to answer the question that was posed with a reasonable degree of precision.

In conclusion, we may say that if the probability that a given trial will produce a specified event cannot be determined on the basis of a rigorous mathematical analysis, the probability can be approximated empirically by performing the trial a vast number of times and counting the number of times the event occurs. The greater the number of trials, the more reliable is the probability thus determined.

11-9. Probability Distribution. A random variable has been defined as one having a value that is influenced by chance, and a trial has been defined as an act having an outcome that is influenced by chance. Therefore, if the outcome of a trial or set of trials is a numerical quantity or expressible in the form of a numerical quantity, this quantity is a random variable.

As an illustration, assume that a shipment of machine parts is to be inspected by examining 20 parts selected at random. If x denotes the number of defective parts among the 20 parts examined, then x is a random variable. Now assume that a box contains green and red balls, and 6 balls are to be drawn at random. If x denotes the number of green balls that are drawn, then x is a random variable. Similarly, assume that a trial has 3 possible outcomes. If the outcome is coded by assigning the number 0 to the first outcome, 10 to the second outcome, and 20 to the third outcome, then the code number is a random variable.

As stated in Art. 10-1, a random variable is discrete if its possible values are discrete, and continuous if its possible values form a continuum.

If a random variable x is discrete and the number of values it can assume is finite, we can assign probability to each possible value of x. The sum of these probabilities must perforce be 1, since it is certain that x will assume one of its possible values and the probability of certainty is 1. The manner in which this total probability of 1 is distributed among the possible values of x is called the "probability distribution" of that variable. For example, assume that a random variable x can have the value 5, 8, or 10, and that the probability associated with each possible value is as follows:

$$P(5) = 0.24 \qquad P(8) = 0.63 \qquad P(10) = 0.13$$

This set of values constitutes the probability distribution of x.

Example 11-11. A bowl contains three chips bearing the numbers -2, -1, and 0. Another bowl contains five chips bearing the numbers 4, 6, 7, 8, and 9. A number will be drawn at random from each bowl. If x denotes the sum of the numbers that are drawn, determine the probability distribution of x.

Solution. The possible values of x and the ways in which x can assume these values are shown in Table 11-3. There are 15 permutations of integers, and the probability distribution is as follows:

TABLE 11-3. POSSIBLE VALUES OF RANDOM VARIABLE

$-2 + 4 = 2$		
	$-1 + 4 = 3$	
$-2 + 6 = 4$		$0 + 4 = 4$
$-2 + 7 = 5$	$-1 + 6 = 5$	
$-2 + 8 = 6$	$-1 + 7 = 6$	$0 + 6 = 6$
$-2 + 9 = 7$	$-1 + 8 = 7$	$0 + 7 = 7$
	$-1 + 9 = 8$	$0 + 8 = 8$
		$0 + 9 = 9$

$$P(2) = \frac{1}{15} \qquad P(3) = \frac{1}{15} \qquad P(4) = \frac{2}{15}$$

$$P(5) = \frac{2}{15} \qquad P(6) = \frac{3}{15} = \frac{1}{5} \qquad P(7) = \frac{3}{15} = \frac{1}{5}$$

$$P(8) = \frac{2}{15} \qquad P(9) = \frac{1}{15}$$

Probability distributions tend to follow certain clearly defined patterns, and we shall discuss one such pattern at this point. Assume that a trial has solely two possible outcomes, which we shall label for convenience as success and failure. Consider that n independent trials will be performed, and let r denote the number of successes that occur. This random variable can assume any integral value from 0 to n, inclusive. The probability distribution of r is referred to as a "binomial distribution." Thus, Theorem 11-7, which expresses the probability corresponding to every possible value of r, is the law of binomial distribution.

If a random variable is continuous, the number of possible values is infinite. Therefore, we cannot properly speak of the probability that the variable will assume some *specific* value, for that probability is zero. The only meaningful statement we can make is one that expresses the probability that the variable will assume a value lying within some specific interval, e.g., between 0 and 5 or between 2.36 and 2.37.

11-10. Probability Histograms. If the probability distribution of a random variable x is depicted graphically, the viewer can instantly perceive the manner in which probability tends to vary with x. As an illustration, we shall apply the data of Example 11-11. The probability distribution is represented in Fig. 11-2, where values of x are plotted on

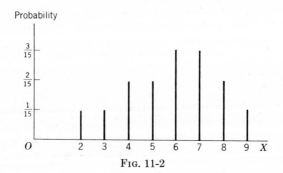

FIG. 11-2

the horizontal axis and probabilities on the vertical axis. The graph reveals the interesting feature that the rate of descent from maximum probability is twice the rate of ascent to that value.

For a discrete variable, the probability corresponding to a particular value can be represented by the length of a line. Analogously, for a continuous variable, the probability corresponding to a particular *interval* can be represented by the area under a graph. To illustrate the manner in which this is done, assume that a continuous random variable x can have any value between -2.0 and 3.0, including these boundary values. Expressed symbolically, $-2.0 \leq x \leq 3.0$. Consider that this range is first divided into the half-unit intervals shown in Fig. 11-3. Now con-

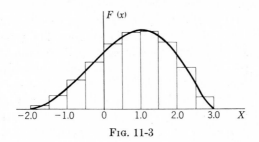

Fig. 11-3

sider that the probability corresponding to each interval is determined, either by mathematical analysis or empirically. A rectangle having that probability as its area is then drawn within that interval and directly above the x axis. For example, assume that the probability that x lies between 1.5 and 2.0 is found to be 0.156. A rectangle is drawn within that interval having a height of 0.156/0.5 or 0.312 vertical units. Now consider that the intervals are made successively smaller. As the width of the intervals approaches zero and the number of intervals therefore increases beyond bound, the diagram approaches the smooth curve shown, which is the probability diagram of the continuous variable.

Assume that we wish to know the probability that $1.0 \leq x \leq 1.65$. As shown in Fig. 11-4, this probability is represented by the area bounded by the curve, the XX axis, and vertical lines at $x = 1.0$ and $x = 1.65$.

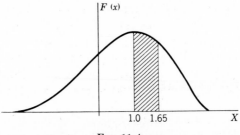

Fig. 11-4

Let $F(x)$ denote the ordinate to the probability diagram. Then, in general,

$$P(a \leq x \leq b) = \int_a^b F(x) \, dx = \text{area under probability curve}$$

A diagram in which values of a variable are plotted on one scale and the relative frequency with which those values occur is plotted on a perpendicular scale is called a "histogram." As the number of trials associated with a random variable increases without limit, relative frequencies approach relative probabilities. Therefore, the probability diagrams in Figs. 11-2 and 11-3 may be regarded as histograms.

The ordinate $F(x)$ to the probability curve of a continuous random variable is referred to as the "probability-density function" of that variable. It is analogous to entropy in thermodynamics. If a thermodynamic process is reversible, the quantity of heat supplied to the substance during the process is represented by the area under the temperature-entropy diagram. Therefore, except for the fact that $F(x)$ is plotted on the vertical axis and entropy is plotted on the horizontal axis, the probability-density function and entropy play a similar role in their respective diagrams: They permit the probability corresponding to a given interval of values or the heat supplied in a reversible thermodynamic process to be represented as the area under a curve.

11-11. Techniques in Calculating Probability. The theorems of Art. 11-7 provide the basic rules for calculating probability. However, while Theorem 11-2 is fundamental, the remaining theorems should be viewed as helpful guides rather than as laws that must be followed mechanically in all situations where they are relevant. In many situations, the crux of the problem is to identify the various ways in which the specified event can occur, and the laws of probability offer no assistance in this matter. Moreover, even in those instances where the probability can readily be calculated by applying a set formula, it is often desirable to test the result by returning to the basic concept of probability and calculating the number of satisfactory outcomes and the number of possible outcomes. In calculating probability, reliance must ultimately be placed on clearly visualizing the problem at hand and solving it by logic. There are no rules or formulas that are tailored to fit every situation.

Example 11-12. In a manufacturing plant, machine A produces 70 per cent and machine B 30 per cent of the units of a standard commodity. The rate of defectives has been found to average 1.5 per cent for machine A and 4.0 per cent for machine B. If a unit is found to be defective, what is the probability (to three decimal places) that it came from machine B?

Solution. Assume that 10,000 units are produced during a given period of time. Of these, 7,000 were produced by machine A and 3,000 by machine B. Applying the average rate of defectives, we obtain the following:

$$\text{Number of defectives produced by A} = 7,000(0.015) = 105$$
$$\text{Number of defectives produced by B} = 3,000(0.040) = 120$$
$$\text{Total number of defectives} = 225$$

$$\text{Probability that defective came from B} = \frac{120}{225} = 0.533$$

Example 11-13. A lot containing 48 units of a standard commodity is to be inspected by examining 6 units selected at random. The lot will be considered acceptable if not more than one of the examined units is defective. What is the probability (to three decimal places) that a lot containing 5 defectives will be rejected?

Solution. Since the selected units are not returned to the lot, each trial is dependent on the preceding trials, and therefore, Theorem 11-7 is not directly applicable. However, the method of solution is similar to that in Example 11-10.

Let x denote the number of examined units that are found to be defective. This variable can have any integral value from 0 to 5, inclusive, and the lot will be accepted only if $x = 0$ or $x = 1$. Therefore, to reduce the number of calculations, we shall determine first the probability that the lot will be accepted.

As in Example 11-10, we establish a hierarchy of events and define a second-order event as one in which the defective and satisfactory units are drawn in some definite sequence. Let D denote a defective unit and S a satisfactory unit. The lot contains 5 defective units and 43 satisfactory units.

The only second-order event corresponding to $x = 0$ is S-S-S-S-S-S. The probability of this event is

$$P(0) = \frac{43}{48} \frac{42}{47} \frac{41}{46} \frac{40}{45} \frac{39}{44} \frac{38}{43} = 0.4968$$

A second-order event corresponding to $x = 1$ is D-S-S-S-S-S. The probability of this event is

$$\frac{5}{48} \frac{43}{47} \frac{42}{46} \frac{41}{45} \frac{40}{44} \frac{39}{43} = 0.0654$$

How many second-order events correspond to $x = 1$? Since D can be assigned to any one of 6 trials, there are 6 such events. Then

$$P(1) = 6(0.0654) = 0.3924$$
$$P(\text{accepted}) = P(0) + P(1) = 0.889$$

Since it is certain that the lot will be either accepted or rejected and the probability of certainty is 1, it follows that

$$P(\text{rejected}) = 1 - 0.889 = 0.111$$

Example 11-14. A box contains 10 similar machine parts. Owing to minor differences among the parts, only 6 of these are suitable for a particular project. A part will be drawn at random from the box, tested, and set aside if it is found to be unsatisfactory. The process will continue until a satisfactory part is found. Let x denote the number of parts that must be tested. Find the probability distribution of x, to four decimal places.

Solution

$$P(1) = \frac{6}{10} = 0.6000$$

$$P(2) = \frac{4}{10}\frac{6}{9} = 0.2667$$

$$P(3) = \frac{4}{10}\frac{3}{9}\frac{6}{8} = 0.1000$$

$$P(4) = \frac{4}{10}\frac{3}{9}\frac{2}{8}\frac{6}{7} = 0.0286$$

$$P(5) = \frac{4}{10}\frac{3}{9}\frac{2}{8}\frac{1}{7}\frac{6}{6} = 0.0048$$

$$\text{Total} \qquad 1.0001$$

It is interesting to observe the relationships among these probabilities, which are as follows:

$$P(2) = \frac{4}{9}P(1) \qquad P(3) = \frac{3}{8}P(2)$$

$$P(4) = \frac{2}{7}P(3) \qquad P(5) = \frac{1}{6}P(4)$$

Example 11-15. A case contains 8 units of a standard commodity. It is known that 3 units are defective and 5 are satisfactory. A unit will be selected at random, examined, and set aside, the process being repeated until all 3 defectives have been identified. Let x denote the number of units that must be examined. Determine the probability distribution of x. (In the special case where 2 defective units have been found and only 1 unit remains in the case, assume for consistency that this unit will also be examined.)

Solution. We shall record first a set of values that will be needed in the solution. By referring to Table 11-2, we obtain the following:

$$P_{2,2(2)} = 1 \qquad P_{3,3(2,1)} = 3 \qquad P_{4,4(2,2)} = 6$$
$$P_{5,5(2,3)} = 10 \qquad P_{6,6(2,4)} = 15 \qquad P_{7,7(2,5)} = 21$$
$$P_{8,8(3,5)} = 56$$

The solution will be based on a study of permutations. Consider the two permutations *acdee* and *adceee*. Since both permutations terminate with three *e*'s, they are distinct from one another solely because the portions to the left of the *e*'s are distinct. Therefore, if both permutations are truncated after the third letter, the two truncated permutations are also distinct.

We now consider the problem at hand. Assume tentatively that the number of defective units is unknown, and therefore, all 8 units will be removed individually and tested to identify the defective ones. Let D denote a defective and S a satisfactory unit. The following are two possible events:

$$S\text{-}D\text{-}S\text{-}S\text{-}D\text{-}D\text{-}S\text{-}S \qquad D\text{-}S\text{-}D\text{-}S\text{-}S\text{-}D\text{-}S\text{-}S$$

Each event constitutes a permutation of three D's and five S's. Therefore, the number of possible events is $P_{8,8(3,5)}$ or 56. Since these events are of equal probability, each event has a probability of 1/56.

In reality, the selection and testing of units terminates when all 3 defective units have been identified. Therefore, the permutations shown above are truncated immediately to the right of the third D. For example, the permutation $S\text{-}D\text{-}S\text{-}S\text{-}D\text{-}D\text{-}S\text{-}S$ becomes $S\text{-}D\text{-}S\text{-}S\text{-}D\text{-}D$. Now, there are either only S's to the right of the cutoff point or no letters whatever. Since the original 56 permutations were all distinct, it follows that the truncated permutations are also all distinct. Therefore, every true event has a probability of 1/56.

To test this conclusion, consider the particular event $D\text{-}S\text{-}S\text{-}D\text{-}S\text{-}D$. Its probability is

$$\frac{3}{8} \frac{5}{7} \frac{4}{6} \frac{2}{5} \frac{3}{4} \frac{1}{3} = \frac{1}{56}$$

Similarly, consider the particular event $S\text{-}D\text{-}D\text{-}D$. Its probability is

$$\frac{5}{8} \frac{3}{7} \frac{2}{6} \frac{1}{5} = \frac{1}{56}$$

We shall determine the probability distribution by analyzing the truncated permutations. Each permutation has three characteristics: it ends in D, it contains a total of three D's, and it contains at most five S's.

Let $x = 3$. The sole permutation that produces this value is D-D-D, and therefore, $P(3) = 1/56$.

Now let $x = 4$. One permutation that produces this value is S-D-D-D. However, the first three letters can be rearranged, and the number of permutations for which $x = 4$ is therefore $P_{3,3(2,1)}$ or 3. Therefore, $P(4) = 3/56$.

Now let $x = 5$. One permutation that produces this value is S-S-D-D-D. However, the first four letters can be rearranged, and the number of permutations for which $x = 5$ is therefore $P_{4,4(2,2)}$ or 6. Therefore, $P(5) = 6/56 = 3/28$.

Continuing these calculations, we obtain the following probability distribution of x:

$$P(3) = \frac{1}{56} \qquad\qquad P(4) = \frac{3}{56}$$

$$P(5) = \frac{6}{56} = \frac{3}{28} \qquad\qquad P(6) = \frac{10}{56} = \frac{5}{28}$$

$$P(7) = \frac{15}{56} \qquad\qquad P(8) = \frac{21}{56} = \frac{3}{8}$$

In general,
$$P(k) = \frac{k^2 - 3k + 2}{2 \cdot 56}$$

The foregoing probabilities total 1, as they should.

11-12. Recurrent Events.

We shall review Example 11-14 for the purpose of constructing a model for our present discussion. In that example, there was a box containing 10 machine parts, only 6 of which were suitable for a particular project. The parts were selected at random and tested until a satisfactory part was found. The basic event, although not explicitly defined in the example, was finding a satisfactory part. The number of trials required to produce the event was denoted by x, and the probability distribution of x was determined.

Now consider that there is a set of such boxes. The boxes are identical insofar as each contains 6 satisfactory and 4 unsatisfactory parts. After the mechanic has found a satisfactory part in one box, he proceeds to the next box and repeats the operation. The specified event (finding a satisfactory part) will occur repeatedly, and therefore, it is referred to as a "recurrent event" in contrast to the isolated event of Example 11-14.

The probability distribution associated with a given box is independent of past occurrences. For example, according to the results of Example 11-14, the probability that the mechanic will find a satisfactory part in the seventh box on the first trial is 0.6000, regardless of how many trials were required for each of the preceding boxes.

We define the interval between two successive events as the number of trials required to produce the later event after the earlier event has occurred. As an illustration, assume that a satisfactory part was found in the fifth box on the third trial. Then the interval between the fourth and fifth events is 3 trials.

Assume that we wish to have the recurrent event occur at constant intervals. For example, if a satisfactory part is to be found on every third trial, the constant interval between successive events is 3 trials. Since the probability distribution for a given box is completely independent of past occurrences, the variable x as defined for the isolated event of Example 11-14 may be redefined for the recurrent event as the interval between successive events. In Example 11-14, it was found that $P(3) = 0.1000$. Therefore, in the present instance, the probability that the interval between successive events will be 3 trials is 0.1000.

In the foregoing situation in which there are identical boxes of machine parts, the interval between successive events is a discrete number of trials, and therefore, it is a discrete random variable. However, there is an important class of recurrent events for which the interval between successive events is a continuum of time or space, and this interval is a continuous rather than a discrete random variable. The following will serve as illustrations:

1. The event is defined as receiving a telephone call. The interval between successive events is the time between two calls.

2. A diver is exploring a lake, and the event is defined as finding a certain type of plant at the bottom of the lake. The interval between successive events is the distance between two points at which the diver located this plant in succession.

If the interval between successive events is a discrete number of trials, the recurrent event is referred to as a trial event. If the interval between successive events is an interval of time or space, the recurrent event is referred to as a time or space event, respectively.

With respect to the situation in which there were identical boxes of machine parts, the probability distribution corresponding to one box was identical for all boxes. Analogously, with respect to a time or space event, it is assumed in the absence of any statement to the contrary that its probability distribution for a given period of time or space is identical for all such periods. To state this in another manner, it is assumed that the probability distribution of the event is uniform throughout time or space, within the time or space encompassed by the situation. For example, if 20 identical machines are in operation and it is found that on the average 3 break down in one month, it is assumed that breakdowns average 3 per month for the entire period of time under consideration.

Assume that on the basis of past experience it is found that telephone calls coming at random average 3 per hour. For our present purpose, this statement will be rephrased by saying that the expected number of calls received in 1 hour is 3. Since the probability distribution is assumed to be uniform, the expected number of calls in 2 hours is 6, and the expected number of calls in 3 hours is 9. Now contracting the time period, we say that the probability that a call will be received in a 10-min period is 3(10/60) or 1/2, and the probability that a call will be received in a 5-min period is 1/4. Similarly, if it is found that flaws in a manufactured pipe occur at random and average 1 in 80 ft of length, the probability of finding a flaw in a 1-ft length of pipe is 1/80.

11-13. Poisson Probability Distribution. We shall now analyze the probability distribution of a time or space event. Consider the following problem. Telephone calls are received at a random rate and average 6 per hour. What is the probability that 5 calls will be received in a 30-min period? Since the event under consideration (receiving a call) can occur at any instant, we may regard each instant of time as a "trial" in the sense that the event either will or will not occur at that instant. However, the number of trials in a 30-min period is infinite.

To determine the probability associated with a time event, we shall employ a technique analogous to that employed in Chap. 9, where we computed the effective interest rate when interest is compounded continuously. First, we divide the time period under consideration into "strips" of definite and uniform length. Each strip of time is regarded as an independent trial. This means that we are assuming that the event either will occur once or not at all within that strip of time, and that the probability that the event will occur is not influenced by the past. Thus, we undertake our analysis with a finite number of trials. Then we approach reality by allowing the length of each strip to become progressively smaller, the process being continued indefinitely. Thus, the length of the strip becomes infinitesimally small and the number of trials becomes infinitely large.   Let

$$T = \text{time period under consideration}$$
$$n = \text{number of time strips in } T$$
$$p = \text{probability specified event will occur in one time strip}$$
$$m = np = \text{expected number of events in period } T$$
$$r = \text{desired number of events in period } T$$

In accordance with Theorem 11-7, the probability that the event will occur in r time strips and fail to occur in the remaining $n - r$ strips is

$$P(r) = \frac{n!}{r!(n - r)!} p^r (1 - p)^{n-r}$$

Now,

$$p^r = \frac{(np)^r}{n^r} = \frac{m^r}{n^r}$$

The foregoing equation can now be recast in this form:

$$P(r) = \frac{n!}{(n - r)!n^r} \frac{m^r}{r!} \frac{(1 - p)^n}{(1 - p)^r}$$

Now consider that the length of the time strip is successively halved while T and r remain constant.   The value of n is successively doubled, the value of p is successively halved, and the value of m remains constant.   Proceeding as in Art. 9-2, we find that as the process continues indefinitely the various terms approach the following limits:

$$\lim \frac{n!}{(n - r)!n^r} = 1$$
$$\lim (1 - p)^n = e^{-pn} = e^{-m} \qquad \lim (1 - p)^r = 1$$

where e is the quantity defined in Art. 9-2.   Thus, in the limit,

$$P(r) = \frac{m^r}{r!} e^{-m} \qquad\qquad (11\text{-}10)$$

In the special case where $r = 0$, Eq. (11-10) reduces to

$$P(0) = e^{-m} \qquad\qquad (11\text{-}10a)$$

Values of e^{-m} can readily be found by the use of natural or common logarithms. For example, to find $e^{-2.46}$ by common logarithms, we have the following:

$$\log e = 0.43429$$
$$\log e^{2.46} = 2.46(0.43429) = 1.06835$$
$$\log e^{-2.46} = -1.06835 = (10 - 1.06835) - 10$$
$$= 8.93165 - 10$$

Then $\qquad\qquad\qquad e^{-2.46} = 0.0854$

It is instructive to investigate the variation of $P(r)$ with r. From Eq. (11-10),

$$\frac{P(r)}{P(r-1)} = \frac{m}{r}$$

Assume that m has an integral value, and consider that r assumes successively higher values. The foregoing equation discloses that $P(r)$ is increasing when $r < m$ and decreasing when $r > m$, and that $P(r)$ has its maximum value when $r = m - 1$ and $r = m$. As r increases without limit, $P(r)$ approaches zero, as we would expect.

Equation (11-10) was first derived by the French mathematician Poisson, and a time or space event to which this equation applies is said to have a Poisson probability distribution.

Example 11-16. A firm manufactures long rolls of tape and then cuts the roll into 1,200-ft lengths. Extensive measurements have shown that defects in the roll occur at random and the average distance between defects is 400 ft.

a. What is the probability (to four decimal places) that a tape has at most 4 defects?

b. What is the probability that a tape has more than 2 defects?

c. If a case contains three tapes, what is the probability that one tape has 4 defects and the remaining tapes have 3 defects each?

Solution. Let x denote the number of defects in the 1,200-ft tape.

$$m = \frac{1,200}{400} = 3 \qquad e^{-3} = 0.04979$$

By Eq. (11-10),

$$P(0) = e^{-3} = 0.04979$$

$$P(1) = \frac{3}{1!} e^{-3} = \frac{3}{1} (0.04979) = 0.14937$$

$$P(2) = \frac{3^2}{2!} e^{-3} = \frac{9}{2} (0.04979) = 0.22406$$

$$P(3) = \frac{3^3}{3!} e^{-3} = \frac{27}{6} (0.04979) = 0.22406$$

$$P(4) = \frac{3^4}{4!} e^{-3} = \frac{81}{24} (0.04979) = 0.16804$$

Part a

$$P(x \leq 4) = P(0) + P(1) + P(2) + P(3) + P(4)$$
$$= 0.8153$$

Part b

$$P(x \leq 2) = P(0) + P(1) + P(2) = 0.4232$$
$$P(x > 2) = 1 - 0.4232 = 0.5768$$

Part c.   Consider that the first tape is to have 4 defects and the second and third tapes are to have 3 defects each.   This event corresponds to the permutation 4-3-3, and its probability is

$$0.16804(0.22406)^2 = 0.00844$$

Three such permutations can be formed, the other two being 3-4-3 and 3-3-4. These three permutations have equal probability, and therefore, the probability of the specified event is

$$3(0.00844) = 0.0253$$

In general, assume that a case contains n tapes.   The probability that m tapes have r_1 defects and the remaining tapes have r_2 defects is

$$\frac{n!}{m!(n-m)!} [P(r_1)]^m \times [P(r_2)]^{n-m}$$

When a time or space event has a Poisson probability distribution, interest often centers about the interval between successive events. Since this interval is a continuous random variable, we cannot express the probability that it will have some specific value; we can only express the probability that its value will lie within a specified interval.

Example 11-17. Experience indicates that a certain apparatus breaks down on an average of once every 24 operating days. If a Poisson probability distribution is assumed, what is the probability that the time between breakdowns will be at least 30 operating days? What is the probability that it will be less than 30 operating days?

Solution. The time between breakdowns is at least 30 days if no breakdowns occur within a period of 30 days. Let

T = given period of time, operating days
x = number of breakdowns during period T
t = desired interval between breakdowns, operating days

Set $T = 30$. The expected number of breakdowns is

$$m = \frac{30}{24} = 1.25$$

$$P(x = 0) = e^{-m} = e^{-1.25} = 0.287$$

Then
$$P(t \geq 30) = 0.287$$
and
$$P(t < 30) = 1 - 0.287 = 0.713$$

In general, if m denotes the expected number of events during a time period T and t denotes the interval between successive events, then

$$P(t \geq T) = e^{-m} \tag{11-11}$$

With respect to a time event, the interval between successive events is often referred to as the "waiting time" of the event. The calculation of waiting time constitutes the basis of queuing theory.

Equation (11-10) for Poisson probability was derived by starting with Theorem 11-7 for binomial probability and then allowing the number of trials to increase without limit. Thus, Poisson probability is the limit of binomial probability as the number of trials becomes infinite. Therefore, although Poisson probability has meaning solely with respect to a time or space event, it can be applied as a reasonable approximation to a trial event if the probability that the event will occur on a single trial is extremely small. In this situation, the number of trials between successive events tends to be extremely large, and the true probability will depart negligibly from the Poisson probability. The motive in applying this approximation is to simplify the arithmetical calculations, as these would otherwise be formidable.

Example 11-18. Experience indicates that the incidence of defective machine parts averages 1 in every 300. Using a Poisson approximation, determine the probability that a case containing 120 parts has no more than 1 defective part.

Solution. Let x denote the number of defective parts.

$$m = \frac{120}{300} = 0.4$$

$$P(0) = e^{-0.4} = 0.6703 \qquad P(1) = 0.4e^{-0.4} = 0.2681$$
$$P(x \leq 1) = P(0) + P(1) = 0.9384$$

11-14. Reliability. The "reliability" of a mechanism is the probability that it will function satisfactorily. If the mechanism is to operate intermittently and briefly, the reliability refers to its performance on a given occasion. If the mechanism is to operate continuously, the reliability refers to its performance for a specified period of time.

With respect to a mechanism that operates intermittently, the event can be defined as either the success or the failure of the mechanism on a given occasion. This is a trial event, and the reliability of the mechanism can be calculated on the basis of the known data by applying the theorems of Art. 11-7.

Example 11-19. A safety mechanism consists of three devices: A, B, and C. The mechanism operates if A and B are both activated, or if C alone is activated. The reliability is 0.90 for A, 0.85 for B, and 0.82 for C. What is the reliability of the mechanism?

Solution. METHOD 1. If A and B are taken as a unit, there are four possible combined outcomes, as follows:

| A and B | Success | Success | Failure | Failure |
C	Success	Failure	Success	Failure

Only the last combined outcome causes failure of the mechanism, and therefore, we shall calculate the probability of this event.

For A and B taken as a unit,
  $P(\text{success}) = 0.90(0.85) = 0.765$
  $P(\text{failure}) = 1 - 0.765 = 0.235$
For C,
  $P(\text{failure}) = 0.18$

For the mechanism,
$P(\text{failure}) = 0.235(0.18) = 0.0423$
$P(\text{success}) = 1 - 0.0423 = 0.9577$

METHOD 2. Success of the mechanism can result from either of two outcomes:

a. Success of C, irrespective of the success or failure of the A-B unit. The probability of this outcome is 0.82.

b. Failure of C and success of the A-B unit. The probability of this outcome is $0.18(0.90)(0.85) = 0.1377$.

Therefore, for the mechanism,
$P(\text{success}) = 0.82 + 0.1377 = 0.9577$

METHOD 3. It is often preferable and sometimes necessary to consider the outcome with respect to the individual devices. Let the subscripts s and f refer, respectively, to success and failure of the device. There are 8 possible combined outcomes, 5 of which produce success of the mechanism. These outcomes and their respective probabilities are as follows:

$$
\begin{array}{lll}
A_s\text{-}B_s\text{-}C_s & 0.90(0.85)(0.82) = & 0.6273 \\
A_s\text{-}B_s\text{-}C_f & 0.90(0.85)(0.18) = & 0.1377 \\
A_s\text{-}B_f\text{-}C_s & 0.90(0.15)(0.82) = & 0.1107 \\
A_f\text{-}B_s\text{-}C_s & 0.10(0.85)(0.82) = & 0.0697 \\
A_f\text{-}B_f\text{-}C_s & 0.10(0.15)(0.82) = & \underline{0.0123} \\
& \text{Total} & 0.9577
\end{array}
$$

Then $P(\text{success}) = 0.9577$

Now we shall analyze the reliability of a mechanism that operates continuously. The event is defined as the abrupt demise of the mechanism. Consider that a mechanism is set in operation. At the instant this fails, a second mechanism of identical type is set in operation. At the instant the second mechanism fails, a third mechanism of identical type is set in motion, etc. Let

T' = time elapsed since first mechanism was set in operation
m' = number of failures during time T'
T = desired life span of mechanism
m = expected number of failures in time T
$R(T)$ = reliability of mechanism for desired life span; i.e., probability a single mechanism will operate for time T

Abrupt failure of a mechanism is a time event. If T' is extremely large, m' may be regarded as the expected number of failures in time T'. Assume that failure of the mechanism has a Poisson probability distribution. Since the probability distribution is assumed to be uniform in time, it follows that

$$\frac{m}{T} = \frac{m'}{T'} \qquad \text{or} \qquad m = \frac{T}{T'} m'$$

By Eq. (11-10a), the probability that there will be no failures during the time T is e^{-m}. Since this probability is the reliability of the mechanism for a life span T, it follows that

$$R(T) = e^{-m} \qquad (11\text{-}12)$$

Since m is directly proportional to T, the foregoing equation can be recast as

$$R(T) = e^{-kT} \qquad (11\text{-}12a)$$

where k is a constant of proportionality. If the possibility exists that the mechanism is initially defective, the last equation becomes

$$R(T) = R_0 e^{-kT} \qquad (11\text{-}12b)$$

where R_0 is the initial reliability; i.e., the probability that the mechanism is satisfactory when first placed in service.

Fig. 11-5

Figure 11-5 is the graph of Eq. (11-12b). At each instant of its service life, the reliability of the mechanism is decreasing at a rate directly proportional to its own magnitude, and the mechanism is said to have a negative-exponential reliability.

Where nothing is stated to the contrary, it is to be understood that $R_0 = 1$.

Example 11-20. The transistor control circuit of an electronic device contains three capacitors, a diode, a potentiometer, six resistors, a solenoid, and two transistors. This device will be used in an aircraft having a mission time of 200 hours. The average number of failures of these components in one million hours of operation in an aircraft are as follows: capacitor, 1.5; diode, 30; potentiometer, 37.5; resistor, 37.5; solenoid, 7.5; transistor, 75. Compute the reliability of the device.

Solution. The device fails when any component fails. The expected number of failures of the device in one million hours of operation is calculated in Table 11-4. Then

$$T' = 1,000,000 \qquad m' = 454.5 \qquad T = 200$$

$$m = \left(\frac{200}{1,000,000}\right) 454.5 = 0.0909$$

$$R(200 \text{ hours}) = e^{-0.0909} = 0.913$$

TABLE 11-4. EXPECTED NUMBER OF FAILURES OF DEVICE IN ONE MILLION HOURS OF OPERATION

Component	Number of failures
Capacitors	$3 \times 1.5 = 4.5$
Diode	$1 \times 30 = 30.0$
Potentiometer	$1 \times 37.5 = 37.5$
Resistors	$6 \times 37.5 = 225.0$
Solenoid	$1 \times 7.5 = 7.5$
Transistors	$2 \times 75 = 150.0$
Total	454.5

11-15. Mean and Standard Deviation of Infinite Set of Values. The arithmetical mean and standard deviation of a discrete random variable are defined in Chap. 10. If the number of values the variable can assume is very small and the number of trials is very large, each possible value of the variable will presumably recur frequently, and the arithmetical mean is calculated by applying the relative frequencies of these values.

Let x_i denote a possible value of x and f_i denote the number of times x assumes that value. Equations (10-1) and (10-5) assume the following forms, respectively:

$$x_m = \frac{\Sigma f_i x_i}{n} \tag{11-13}$$

$$\sigma = \sqrt{\frac{\Sigma f_i (x_i - x_m)^2}{n}} \tag{11-14}$$

Example 11-21. An examination consisting of 4 problems was administered to a group of 120 adults. Each individual was given a score equal to the number of problems he solved correctly. The frequency of each score was as follows:

Score	0	1	2	3	4
Frequency	3	27	43	39	8

Calculate the mean score.

Solution

$$x_m = \frac{3 \times 0 + 27 \times 1 + 43 \times 2 + 39 \times 3 + 8 \times 4}{3 + 27 + 43 + 39 + 8} = \frac{262}{120} = 2.18$$

If the probability distribution of a random variable has been established by mathematical analysis, it is anticipated that as the number of trials increases beyond bound the relative frequencies of the possible values will approach the calculated probabilities as limiting values. Therefore,

$$\lim_{n \to \infty} \frac{f_i}{n} = P(x_i) \tag{11-15}$$

Then
$$x_m = \Sigma[P(x_i)]x_i \tag{11-16}$$

and
$$\sigma = \sqrt{\Sigma[P(x_i)](x_i - x_m)^2} \tag{11-17}$$

The arithmetical mean of an infinite set of values as calculated by Eq. (11-16) is referred to as the "expected value" resulting from a single trial.

Example 11-22. With reference to Example 11-11, assume that the pair of chips is returned to the bowl after being drawn and that the act of drawing a pair of chips is repeated indefinitely. Find the arithmetical mean and standard deviation of the resulting set of values of x.
Solution. Refer to Table 11-5.

$$x_m = \frac{87}{15} = 5.800 \qquad \sigma = \sqrt{\frac{54.40}{15}} = 1.904$$

Consider that a trial can have solely two possible outcomes: success or failure. Also consider that n independent trials are performed, and let x denote the number of trials that culminate in success. The value of x can range from 0 to n, inclusive, and the probability associated with each possible value is given by Theorem 11-7. As stated in Art. 11-9, the variable x is said to have a binomial probability distribution. We now visualize the n independent trials that created a value of x as a *set* of trials, and we consider that the set of trials will be repeated indefinitely. This procedure generates an infinite set of values of x.

TABLE 11-5

x_i	$15P(x_i)$	$15[P(x_i)]x_i$	$15[P(x_i)](x_i - x_m)^2$
2	1	2	$1(-3.8)^2 = 14.44$
3	1	3	$1(-2.8)^2 = 7.84$
4	2	8	$2(-1.8)^2 = 6.48$
5	2	10	$2(-0.8)^2 = 1.28$
6	3	18	$3(0.2)^2 = 0.12$
7	3	21	$3(1.2)^2 = 4.32$
8	2	16	$2(2.2)^2 = 9.68$
9	1	9	$1(3.2)^2 = 10.24$
Total		87	54.40

Let p denote the probability of success in an individual trial. By applying Theorem 11-7, it can be shown that the arithmetical mean and standard deviation of this infinite set of values of x are as follows:

$$x_m = np \qquad \sigma = \sqrt{np(1 - p)} \qquad (11\text{-}18)$$

The expression for x_m may be regarded as self-evident.

Example 11-23. A firm manufactures units of a standard commodity. The probability that a unit is defective is 4 per cent. The units are shipped in sets of 300. If x denotes the number of defective units in a shipment, what is the mean value and standard deviation of x in the long run?

Solution

$$n = 300 \qquad p = 0.04$$
$$x_m = np = 12 \qquad \sigma = \sqrt{12(0.96)} = 3.39$$

With respect to a continuous random variable, the properties of its values resulting from an infinite number of trials can be given directly in terms of the properties of the area under its probability curve. It will be recalled that the centroidal axis of an area in a specified direction serves to measure the "average" distance of the area from a line parallel to that axis, and the radius of gyration of the area with respect to its centroidal axis is an index of the dispersion of the area about that axis.

Let x denote a continuous random variable and let A denote the area between the probability curve of x and the horizontal axis. Applying the precise definitions of centroidal axis and radius of gyration, it is seen that:

1. The arithmetical mean of x is the value of x at which the vertical centroidal axis through A is located.

2. The standard deviation of x is the radius of gyration of A with respect to its vertical centroidal axis.

11-16. Normal Probability Distribution. A continuous random variable is said to have a "normal" or "gaussian" probability distribution if the range of its possible values is of infinite extent and the equation of its probability curve has the following form:

$$F(x) = \frac{1}{b\sqrt{2\pi}}\, e^{-(x-a)^2/2b^2} \tag{11-19}$$

where $F(x)$ is the probability-density function and a and b are constants. Equation (11-19) was first developed by De Moivre in solving a problem in gambling, but it was later found that this equation applies to a vast number of random variables associated with natural phenomena. For example, many characteristics of a species, such as height, weight, and intelligence, are found to have a normal probability distribution.

Figure 11-6 is a graph of Eq. (11-19) for assumed values of a and b. This bell-shaped curve has a summit at $x = a$, and it is symmetrical about the vertical line through the summit. Thus, the constant a in Eq. (11-19) is the arithmetical mean of x. It can also be shown that

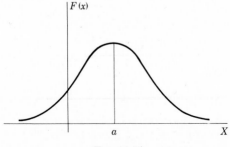

Fig. 11-6

b in this equation is the standard deviation.   The smaller the standard deviation, the steeper the curve.

The determination of probability when the distribution is normal can be expedited by referring to Table C-1 in the Appendix.   This table records the value of the area under the probability curve for an interval that has one boundary at the centerline.   The width of the interval is

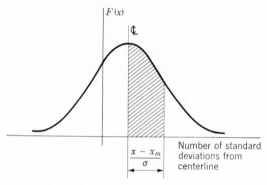

Fɪɢ. 11-7

expressed by using the standard deviation as a unit, as shown in Fig. 11-7.   Let z denote this width.   Then

$$z = \frac{x - x_m}{\sigma} \quad \text{or} \quad z = \frac{x_m - x}{\sigma}$$

Example 11-24.   A continuous random variable x having normal probability distribution is known to have a mean value of 14 and a standard deviation of 2.5. What is the probability that (a) x lies between 14 and 17; (b) x lies between 12 and 16.2; and (c) x is less than 10?

Solution.   Let A denote the area under the probability diagram from the centerline to the vertical line at the given value of x.

$$x_m = 14 \qquad \sigma = 2.5$$

Part a

$$z = \frac{17 - 14}{2.5} = 1.20$$

From Table C-1, $A = 0.38493$
Then $P(14 < x < 17) = 0.38493$

The probability is the same whether the boundary values 14 and 17 are included or excluded.

Part b. Resolve the interval into two parts by cutting it at x_m. For the first part,

$$z = \frac{14 - 12}{2.5} = 0.8 \qquad A = 0.28814$$

For the second part,

$$z = \frac{16.2 - 14}{2.5} = 0.88 \qquad A = 0.31057$$

Then $\qquad P(12 < x < 16.2) = 0.28814 + 0.31057 = 0.59871$

Part c. First take the interval from $x = 10$ to $x = 14$.

$$z = \frac{4}{2.5} = 1.6 \qquad A = 0.44520$$

Then $\qquad P(10 < x < 14) = 0.44520$
$$P(x < 10) = 0.5 - 0.44520 = 0.05480$$

In advanced texts, it is demonstrated that the binomial probability distribution approaches the normal distribution as the number of trials in a set of trials becomes infinitely large. Similarly, the Poisson probability distribution approaches the normal distribution as the expected number of events in a unit of time or space becomes very large. Consequently, in determining the probability of a random variable having a binomial or Poisson distribution, the normal distribution is often applied as an approximation in order to simplify the calculations.

Example 11-25. With reference to Example 11-23, what is the probability that a shipment of 300 units contains less than 9 defective units? Apply a normal distribution to approximate the answer.

Solution

$$x_m = 12 \qquad \sigma = 3.39$$
$$z = \frac{12 - 9}{3.39} = 0.885$$

By linear interpolation,

$$A = 0.31192$$
$$P(x < 9) = 0.5 - 0.31192 = 0.18808$$

Another important type of probability distribution that finds wide application in engineering, economics, and other fields is the lognormal distribution. This is one in which the logarithm of the random variable has a normal probability distribution.

11-17. Economy Analysis with Contingencies. Every business venture is fraught with uncertainty, and consequently the venture includes many items of income and cost that are potential rather than certain. For example, the by-products of a manufacturing process *may* have commercial value, a parcel of land *may* appreciate in value as a result of commercial expansion, and the cost of delivering merchandise *may* increase as a result of increased traffic congestion. Therefore, when alternative proposals are to be compared, it is essential that the economy analysis encompass all contingent income and costs that can reasonably be anticipated.

Contingent income and costs are evaluated by assigning probability to each item. If there is a probability p that an individual will receive a sum of money M, his "expected income" is pM. Similarly, if there is a probability p that an individual will be required to pay a sum of money M, his "expected cost" is pM.

Example 11-26. A maintenance company sends a mechanic to the plant of a customer to install a new meter. On each assignment, the mechanic takes only the meter required for that assignment. He procures the meter from the warehouse of the firm, but practical difficulties preclude his testing the meter in advance. The incidence of defective meters is considered to be 9 per cent; i.e., there are 9 defective meters in every set of 100.

If the meter proves to be defective, the mechanic must return to the warehouse for a replacement. The cost of the return trip and the loss of good will incurred by the firm are assigned a total monetary value of $40. Therefore, the firm is considering the feasibility of having the mechanic take extra meters on each assignment. However, if the mechanic returns unused meters, there is an additional cost in handling, record-keeping, and potential damage; this cost is assigned a monetary value of $3 per meter. The unused meters are returned to the warehouse without being tested.

On the basis of the foregoing data, determine whether it is most economical to have the mechanic take one, two, or three meters on each assignment.

Solution. Let N denote the number of meters taken per assignment and x denote the number of these meters that are defective. We shall refer to the condition in which all N meters are defective as an "inadequacy."

TABLE 11-6

N	x	Probability	Number of inadequacies	Average number of meters returned
1	0	0.9100	0	0
	1	0.0900	1	0
2	0	0.8273	0	1
	1	0.1654	0	0.5
	2	0.0073	1	0
3	0	0.7513	0	2
	1	0.2279	0	1.67
	2	0.0203	0	1
	3	0.0005	1	0

Table 11-6 presents the probability, number of inadequacies, and average number of meters returned corresponding to every set of values of N and x. The probabilities are calculated by following the same procedure as in Example 11-13. For example, if three meters are taken, the probability that one of the three is defective is

$$P(1) = \frac{3!}{1!2!} \frac{9}{100} \frac{91}{99} \frac{90}{98} = 0.2279$$

The average number of meters returned is calculated by considering the various ways in which the mechanic can select the meters he has taken on the assignment. As an illustration, assume that three meters are taken and one is defective. Let D denote the defective meter and S_1 and S_2 denote the satisfactory meters. The six permutations of these three items are as follows:

Permutation	Number returned
D-S_1-S_2	1
D-S_2-S_1	1
S_1-D-S_2	2
S_1-S_2-D	2
S_2-D-S_1	2
S_2-S_1-D	2
Total	10

The process of selecting meters terminates when a meter is found to be satisfactory. Therefore, each permutation terminates when S_1 or S_2 is reached. If the mechanic first selects D and then S_1, he discards D and returns S_2. If he first selects S_1, he returns D and S_2, unaware that one of the two is defective. Since these permutations are of equal probability, we divide the total in the column "Number returned" by the number of permutations, giving 1.67 as the average number of meters returned.

The expected number of inadequacies and the expected number of meters returned corresponding to the three alternative values of N are calculated in Table 11-7. The expected cost is as follows:

$$N = 1: (0.0900)40 \qquad\qquad = \$3.60$$
$$N = 2: (0.0073)40 + (0.9100)3 = \$3.02$$
$$N = 3: (0.0005)40 + (1.9035)3 = \$5.73$$

TABLE 11-7

N	Expected number of inadequacies	Expected number of meters returned
1	$(0.0900)1 = 0.0900$	
2	$(0.0073)1 = 0.0073$	$(0.8273)1 \quad = 0.8273$ $(0.1654)0.5 = 0.0827$ Total $\qquad\quad \overline{0.9100}$
3	$(0.0005)1 = 0.0005$	$(0.7513)2 \quad\; = 1.5026$ $(0.2279)1.67 = 0.3806$ $(0.0203)1 \quad\; = 0.0203$ Total $\qquad\qquad \overline{1.9035}$

The expected cost is minimum when the mechanic takes two meters on each assignment.

As the foregoing calculations disclose, the probability of an inadequacy is slight when $N = 2$. Therefore, increasing N to 3 causes a negligible decrease in the expected cost of an inadequacy but a sharp increase in the expected cost of returning unused meters.

PROBLEMS

11-1. How many permutations can be formed using the letters of the word *regress*, taken all at a time? *Ans.* 630

11-2. Permutations are to be formed of the first 7 letters of the alphabet, taken 4 at a time, with the restriction that d cannot be placed anywhere to the left of c. For example, the permutation *edgc* is unacceptable. How many permutations can be formed? *Ans.* 720

11-3. Group A contains 9 distinct objects and group B contains 6 distinct objects. Permutations of 5 objects are to be formed by selecting 2 objects from group A and 3 objects from group B. How many permutations can be formed? *Ans.* 86,400

11-4. With reference to Prob. 11-3, how many permutations can be formed if 2 of the 6 objects in group B are alike? *Ans.* 51,840

11-5. How many permutations can be formed using the letters of the word *planning*, taken 5 at a time? (The letter *n* can be repeated. For example, *apnin* and *inlnn* are acceptable permutations.) *Ans.* 1,520

Hint: Find the number of permutations that can be formed with no *n*'s, with one *n*, with two *n*'s, and with three *n*'s.

11-6. A business firm has three positions to be filled. There are 3 qualified applicants for the first position, 2 qualified applicants for the second position, and 4 qualified applicants for the third position. In addition, there are 2 applicants who are qualified for all three positions. In how many ways can the three positions be filled? *Ans.* 94

11-7. A committee consisting of 5 members is to be formed, and 12 individuals are available for assignment to the committee. In how many ways can the committee be formed if (a) there are no restrictions, and (b) Smith and Jones cannot both serve on the committee? *Ans.* 792; 672

11-8. A committee consisting of 6 members is to be formed, and 15 individuals are available for assignment. In how many ways can the committee be formed if Wilson will serve only if Adams is also on the committee? *Ans.* 3,718

11-9. A committee will consist of a moderator, a co-moderator, and 3 other members of equal rank. If 7 individuals are available for assignment, in how many ways can the committee be formed? *Ans.* 420

11-10. A box contains 5 red, 3 green, and 2 blue balls. Another box contains 6 red and 4 green balls. A ball will be drawn at random from the first box and placed in the second box. Then a ball will be drawn at random from the second box. What is the probability that (a) a red ball will be drawn both times; (b) the first ball drawn will be red or green and the second ball drawn will be red? *Ans.* 7/22; 53/110

11-11. A box contains 20 balls; 8 are red and 7 are blue. A ball will be drawn at random from the box and then returned to the box, the process being performed 9 times. What is the probability that (a) a red ball will be drawn exactly 3 times; (b) a red ball will be drawn at least 3 times; (c) a red ball will be drawn exactly 3 times and a blue ball will be drawn exactly 4 times? *Ans.* 0.251; 0.768; 0.075

11-12. A box contains 10 bolts. Of these, 4 have square heads and 6 have hexagonal heads. Seven bolts will be removed from the box individually, in a random selection. What is the probability that (a) the first bolt will have a square head and the second bolt a hexagonal head; (b) the fourth bolt will have a square head and the sixth bolt a hexagonal head; (c) the second and third bolts will both have square heads and the fifth bolt a hexagonal head? *Ans.* 4/15; 4/15; 1/10

Hint for Parts b and c: One method of solution consists of forming permutations, each permutation corresponding to the order in which the bolts are drawn. Assume that all 10 bolts will be removed. How many permutations can be formed if no restrictions are imposed? How many permutations can be formed under the imposed restrictions? Explain why Parts *a* and *b* have an identical answer.

11-13. A box contains 12 machine parts, only 8 of which are suitable for a particular project. Parts will be selected at random and tested until 2 satisfactory parts are found. If x denotes the number of parts that must be tested, find the probability distribution of x.

Ans. $P(2) = 0.424$; $P(3) = 0.339$; $P(4) = 0.170$; $P(5) = 0.057$; $P(6) = 0.010$

11-14. A box contains 5 type A units, 7 type B units, and 3 type C units. If 7 units are selected at random without being returned to the box, what is the probability that these consist of 3 type A units, 3 type B units, and 1 type C unit?

Ans. 70/429

11-15. A box contains 5 type A units and 3 type B units. To fill a sales order, it is necessary to take 4 type A units and 1 type B unit from this box. However, because the units were not properly labeled, it will be necessary to select a unit at random and examine it to establish its identity, continuing the process until the required set of units is obtained. If x denotes the number of units that must be examined, determine the probability distribution of x.

Ans. $P(5) = 15/56$; $P(6) = 21/56$; $P(7) = 5/14$

Hint: Consider that $x = 6$. Six trials will yield the required set if after the first 5 trials 1 type A or 1 type B unit is lacking, and the sixth trial fills the gap. Thus, two events that correspond to $x = 6$ are A-A-A-B-B-A and A-A-A-A-A-B. How many sets of 5 trials will yield 3 type A and 2 type B units? How many will yield 5 type A units?

11-16. A typesetting machine is found to produce flaws randomly and on the average of 1 in every 900 words. An article to be set on this machine consists of 7,200 words. What is the probability that the printed article will contain more than 6 flaws? Use a Poisson approximation. *Ans.* 0.689

11-17. In a city of 200,000 families, it is known that 4,000 families use product A. If 150 families are selected at random, what is the probability that more than 4 use product A? Use a Poisson approximation. *Ans.* 0.1847

11-18. A firm received an order for 1,000 units of a standard commodity, and it produced 600 units on machine A and 400 units on machine B. The incidence of defectives is 1 in 300 for machine A and 1 in 400 for machine B. What is the probability that the 1,000 units contain no more than 2 defectives? Use a Poisson approximation. *Ans.* 0.4232

11-19. A safety mechanism consists of three devices: A, B, and C. The mechanism operates if A operates and, in addition, either B or C operates. The reliability is 0.96 for A, 0.92 for B, and 0.75 for C. What is the reliability of the mechanism? *Ans.* 0.9408

11-20. A safety mechanism consists of four devices: A, B, C, and D, arranged in that order. The mechanism operates if any two adjacent devices are activated; i.e., if A and B are both activated, if B and C are both activated, or if C and D are both activated. The reliability is 0.85 for A, 0.82 for B, 0.75 for C, and 0.90 for D. What is the reliability of the mechanism? *Ans.* 0.9108

11-21. A mechanism that is assumed to have a negative-exponential reliability fails on an average of once in every 500 hours of operation. What is the probability that the mechanism will function for (a) 400 hours; (b) 600 hours?

Ans. 0.4493; 0.3012

11-22. A mechanism consists of four components: A, B, C, and D. The mechanism operates if any three components operate. The average number of failures of these components in 10,000 hours of operation is as follows: A, 5; B, 10; C, 3; D, 6. What is the reliability of the mechanism if it is to operate 150 hours? Assume each component has a negative-exponential reliability. *Ans.* 0.9628

11-23. The reliability of a production facility is 0.50 for 100 operating days and 0.18 for 250 operating days. If the reliability varies exponentially with operating time, what is the initial reliability? How frequently does the facility break down on the average (to the nearest integral number of days)?

Ans. 0.988; once every 147 operating days

11-24. A firm produces units of a standard commodity in batches of 500. The probability that a unit is defective is 3 per cent. If x denotes the number of defective units in a batch, what is the mean value and standard deviation of x in the long run?

Ans. 15; 3.81

11-25. The time required to perform a certain operation is assumed to have a normal distribution. Studies of past performance show that the average time required is 5.8 hours and the standard deviation is 0.5 hours. What is the probability that the operation can be performed within 5 hours? *Ans.* 0.0548

11-26. A random variable x having a normal probability distribution is found to have a mean value of 20. What is the probability that x is less than 18 if the standard deviation is (a) 2; (b) 4? Explain why the probability is higher in the second case than in the first case. *Ans.* 0.15866; 0.30854

11-27. A random variable x has a normal probability distribution and a mean value of 30. If $P(x < 27) = 0.2630$, what is $P(x > 32)$? *Ans.* 0.3362

11-28. The following game of chance has been devised: Six \$1 bills, one \$5 bill, and one \$10 bill are placed in a box. An individual is to draw a bill from the box at random, continuing to do so until he has either three \$1 bills or the \$10 bill. He will retain the money drawn. Thus, the individual's income from this game of chance can range from \$3 (if he draws only three \$1 bills) to \$17 (if he draws two \$1 bills, the \$5 bill, and then the \$10 bill). What is his expected income, to the nearest cent? Find the answer by each of the following methods:

METHOD 1. Record all possible outcomes and calculate their respective probabilities.

METHOD 2. Record all permutations of the following numbers, taken all at a time: six 1's, one 5, and one 10. Then truncate each permutation at the point where the specified conditions are satisfied, and find the sum of the numbers in each truncated permutation. *Ans.* \$8.04

11-29. A contractor performs a standard type of work. The time required for completion is assumed to have a normal probability distribution, the average time of completion is 35 working days, and the standard deviation is 2.5 working days. On a particular project, the contractor is offered a bonus of \$500 if the work is completed within 32 days but is required to pay a penalty of $200 if the work is not completed within 40 days. Assuming that the provision for a bonus or penalty has no effect on the time of completion, what is the expected value of the bonus or penalty?

Ans. Bonus of $53

11-30. Devise an alternative proof of Eq. (11-8).

Solution. Using numerical values, consider that a group contains 10 items and that combinations of these items taken 7 at a time are to be formed. Arbitrarily selecting the fourth item, we may consider the combinations to be divided into two sets: those that contain the fourth item and those that do not. The first set of combinations is formed by combining the fourth item with 6 of the remaining 9 items, and the number of combinations in this set is $C_{9,6}$. The second set of combinations is formed by excluding the fourth item from consideration and combining the remaining 9 items 7 at a time. The number of combinations in this set is $C_{9,7}$. Thus,

$$C_{10,7} = C_{9,6} + C_{9,7}$$

11-31. Prove the following:

$$C_{n,0} - C_{n,1} + C_{n,2} - C_{n,3} + \cdots + (-1)^n C_{n,n} = 0$$

Hint: Refer to the proof of Eq. (11-6).

11-32. Consider that the entire set of permutations of n items taken all at a time has been formed. One permutation in this set may be regarded as the "standard" permutation. The position that an item occupies in the standard permutation is called its "standard position." An "aberrant permutation" is one in which all items are in nonstandard position. For example, if *abcd* is the standard permutation of the first 4 letters, then *cadb* is an aberrant permutation.

Let A_n denote the number of aberrant permutations of n items taken all at a time. It can be demonstrated that

$$A_n = nA_{n-1} + (-1)^n$$

or

$$A_n = \frac{n!}{0!} - \frac{n!}{1!} + \frac{n!}{2!} - \frac{n!}{3!} + \cdots + (-1)^n \frac{n!}{n!}$$

If *abcde* is the standard permutation of the first 5 letters of the alphabet, in how many ways can these letters be arranged if (*a*) each letter must be placed in non-standard position; (*b*) each letter must be placed in nonstandard position and *d* is restricted to the third position?

Ans. 44; 11

11-33. Seven individuals have formed a club for the purpose of sharing their books. Each member donates a book, and the books are distributed among the members so that each member receives a book other than the one he donated. If all the books are different, in how many ways can the 7 books be distributed?

Ans. 1,854

APPENDIXES

APPENDIX A

TABLE A-1. COMMON LOGARITHMS OF NUMBERS

N	0	1	2	3	4	5	6	7	8	9
10	0000	0043	0086	0128	0170	0212	0253	0294	0334	0374
11	0414	0453	0492	0531	0569	0607	0645	0682	0719	0755
12	0792	0828	0864	0899	0934	0969	1004	1038	1072	1106
13	1139	1173	1206	1239	1271	1303	1335	1367	1399	1430
14	1461	1492	1523	1553	1584	1614	1644	1673	1703	1732
15	1761	1790	1818	1847	1875	1903	1931	1959	1987	2014
16	2041	2068	2095	2122	2148	2175	2201	2227	2253	2279
17	2304	2330	2355	2380	2405	2430	2455	2480	2504	2529
18	2553	2577	2601	2625	2648	2672	2695	2718	2742	2765
19	2788	2810	2833	2856	2878	2900	2923	2945	2967	2989
20	3010	3032	3054	3075	3096	3118	3139	3160	3181	3201
21	3222	3243	3263	3284	3304	3324	3345	3365	3385	3404
22	3424	3444	3464	3483	3502	3522	3541	3560	3579	3598
23	3617	3636	3655	3674	3692	3711	3729	3747	3766	3784
24	3802	3820	3838	3856	3874	3892	3909	3927	3945	3962
25	3979	3997	4014	4031	4048	4065	4082	4099	4116	4133
26	4150	4166	4183	4200	4216	4232	4249	4265	4281	4298
27	4314	4330	4346	4362	4378	4393	4409	4425	4440	4456
28	4472	4487	4502	4518	4533	4548	4564	4579	4594	4609
29	4624	4639	4654	4669	4683	4698	4713	4728	4742	4757
30	4771	4786	4800	4814	4829	4843	4857	4871	4886	4900
31	4914	4928	4942	4955	4969	4983	4997	5011	5024	5038
32	5051	5065	5079	5092	5105	5119	5132	5145	5159	5172
33	5185	5198	5211	5224	5237	5250	5263	5276	5289	5302
34	5315	5328	5340	5353	5366	5378	5391	5403	5416	5428
35	5441	5453	5465	5478	5490	5502	5514	5527	5539	5551
36	5563	5575	5587	5599	5611	5623	5635	5647	5658	5670
37	5682	5694	5705	5717	5729	5740	5752	5763	5775	5786
38	5798	5809	5821	5832	5843	5855	5866	5877	5888	5899
39	5911	5922	5933	5944	5955	5966	5977	5988	5999	6010
40	6021	6031	6042	6053	6064	6075	6085	6096	6107	6117
41	6128	6138	6149	6160	6170	6180	6191	6201	6212	6222
42	6232	6243	6253	6263	6274	6284	6294	6304	6314	6325
43	6335	6345	6355	6365	6375	6385	6395	6405	6415	6425
44	6435	6444	6454	6464	6474	6484	6493	6503	6513	6522
45	6532	6542	6551	6561	6571	6580	6590	6599	6609	6618
46	6628	6637	6646	6656	6665	6675	6684	6693	6702	6712
47	6721	6730	6739	6749	6758	6767	6776	6785	6794	6803
48	6812	6821	6830	6839	6848	6857	6866	6875	6884	6893
49	6902	6911	6920	6928	6937	6946	6955	6964	6972	6981
50	6990	6998	7007	7016	7024	7033	7042	7050	7059	7067
51	7076	7084	7093	7101	7110	7118	7126	7135	7143	7152
52	7160	7168	7177	7185	7193	7202	7210	7218	7226	7235
53	7243	7251	7259	7267	7275	7284	7292	7300	7308	7316
54	7324	7332	7340	7348	7356	7364	7372	7380	7388	7396

N	0	1	2	3	4	5	6	7	8	9
55	7404	7412	7419	7427	7435	7443	7451	7459	7466	7474
56	7482	7490	7497	7505	7513	7520	7528	7536	7543	7551
57	7559	7566	7574	7582	7589	7597	7604	7612	7619	7627
58	7634	7642	7649	7657	7664	7672	7679	7686	7694	7701
59	7709	7716	7723	7731	7738	7745	7752	7760	7767	7774
60	7782	7789	7796	7803	7810	7818	7825	7832	7839	7846
61	7853	7860	7868	7875	7882	7889	7896	7903	7910	7917
62	7924	7931	7938	7945	7952	7959	7966	7973	7980	7987
63	7993	8000	8007	8014	8021	8028	8035	8041	8048	8055
64	8062	8069	8075	8082	8089	8096	8102	8109	8116	8122
65	8129	8136	8142	8149	8156	8162	8169	8176	8182	8189
66	8195	8202	8209	8215	8222	8228	8235	8241	8248	8254
67	8261	8267	8274	8280	8287	8293	8299	8306	8312	8319
68	8325	8331	8338	8344	8351	8357	8363	8370	8376	8382
69	8388	8395	8401	8407	8414	8420	8426	8432	8439	8445
70	8451	8457	8463	8470	8476	8482	8488	8494	8500	8506
71	8513	8519	8525	8531	8537	8543	8549	8555	8561	8567
72	8573	8579	8585	8591	8597	8603	8609	8615	8621	8627
73	8633	8639	8645	8651	8657	8663	8669	8675	8681	8686
74	8692	8698	8704	8710	8716	8722	8727	8733	8739	8745
75	8751	8756	8762	8768	8774	8779	8785	8791	8797	8802
76	8808	8814	8820	8825	8831	8837	8842	8848	8854	8859
77	8865	8871	8876	8882	8887	8893	8899	8904	8910	8915
78	8921	8927	8932	8938	8943	8949	8954	8960	8965	8971
79	8976	8982	8987	8993	8998	9004	9009	9015	9020	9025
80	9031	9036	9042	9047	9053	9058	9063	9069	9074	9079
81	9085	9090	9096	9101	9106	9112	9117	9122	9128	9133
82	9138	9143	9149	9154	9159	9165	9170	9175	9180	9186
83	9191	9196	9201	9206	9212	9217	9222	9227	9232	9238
84	9243	9248	9253	9258	9263	9269	9274	9279	9284	9289
85	9294	9299	9304	9309	9315	9320	9325	9330	9335	9340
86	9345	9350	9355	9360	9365	9370	9375	9380	9385	9390
87	9395	9400	9405	9410	9415	9420	9425	9430	9435	9440
88	9445	9450	9455	9460	9465	9469	9474	9479	9484	9489
89	9494	9499	9504	9509	9513	9518	9523	9528	9533	9538
90	9542	9547	9552	9557	9562	9566	9571	9576	9581	9586
91	9590	9595	9600	9605	9609	9614	9619	9624	9628	9633
92	9638	9643	9647	9652	9657	9661	9666	9671	9675	9680
93	9685	9689	9694	9699	9703	9708	9713	9717	9722	9727
94	9731	9736	9741	9745	9750	9754	9759	9763	9768	9773
95	9777	9782	9786	9791	9795	9800	9805	9809	9814	9818
96	9823	9827	9832	9836	9841	9845	9850	9854	9859	9863
97	9868	9872	9877	9881	9886	9890	9894	9899	9903	9908
98	9912	9917	9921	9926	9930	9934	9939	9943	9948	9952
99	9956	9961	9965	9969	9974	9978	9983	9987	9991	9996

APPENDIX B. COMPOUND INTEREST FACTORS

DEFINITION OF SYMBOLS

1. $s_{\bar{n}}$ = value of sinking fund at terminal date, where periodic deposit is \$1

$$= \frac{(1 + i)^n - 1}{i}$$

2. R'_n = uniform periodic deposit of a sinking fund, where value of fund at terminal date is \$1

$$= \frac{1}{s_{\bar{n}}} = \frac{i}{(1 + i)^n - 1}$$

3. $_nt_0$ = value of sinking fund at origin date, where periodic deposit is \$1

$$= \frac{1 - (1 + i)^{-n}}{i}$$

4. R''_n = uniform periodic deposit of a sinking fund, where value of fund at origin date is \$1

$$= \frac{1}{_nt_0} = \frac{i}{1 - (1 + i)^{-n}}$$

n	$(1+i)^n$	$(1+i)^{-n}$	$s_{\bar{n}}$	R'_n	$_n t_0$	R''_n
1	1.01000	0.99010	1.00000	1.00000	0.99010	1.01000
2	1.02010	0.98030	2.01000	0.49751	1.97040	0.50751
3	1.03030	0.97059	3.03010	0.33002	2.94099	0.34002
4	1.04060	0.96098	4.06040	0.24628	3.90197	0.25628
5	1.05101	0.95147	5.10101	0.19604	4.85343	0.20604
6	1.06152	0.94205	6.15202	0.16255	5.79548	0.17255
7	1.07214	0.93272	7.21354	0.13863	6.72819	0.14863
8	1.08286	0.92348	8.28567	0.12069	7.65168	0.13069
9	1.09369	0.91434	9.36853	0.10674	8.56602	0.11674
10	1.10462	0.90529	10.46221	0.09558	9.47130	0.10558
11	1.11567	0.89632	11.56683	0.08645	10.36763	0.09645
12	1.12683	0.88745	12.68250	0.07885	11.25508	0.08885
13	1.13809	0.87866	13.80933	0.07241	12.13374	0.08241
14	1.14947	0.86996	14.94742	0.06690	13.00370	0.07690
15	1.16097	0.86135	16.09690	0.06212	13.86505	0.07212
16	1.17258	0.85282	17.25786	0.05794	14.71787	0.06794
17	1.18430	0.84438	18.43044	0.05426	15.56225	0.06426
18	1.19615	0.83602	19.61475	0.05098	16.39827	0.06098
19	1.20811	0.82774	20.81090	0.04805	17.22601	0.05805
20	1.22019	0.81954	22.01900	0.04542	18.04555	0.05542
21	1.23239	0.81143	23.23919	0.04303	18.85698	0.05303
22	1.24472	0.80340	24.47159	0.04086	19.66038	0.05086
23	1.25716	0.79544	25.71630	0.03889	20.45582	0.04889
24	1.26973	0.78757	26.97346	0.03707	21.24339	0.04707
25	1.28243	0.77977	28.24320	0.03541	22.02316	0.04541
26	1.29526	0.77205	29.52563	0.03387	22.79520	0.04387
27	1.30821	0.76440	30.82089	0.03245	23.55961	0.04245
28	1.32129	0.75684	32.12910	0.03112	24.31644	0.04112
29	1.33450	0.74934	33.45039	0.02990	25.06579	0.03990
30	1.34785	0.74192	34.78489	0.02875	25.80771	0.03875
35	1.41660	0.70591	41.66028	0.02400	29.40858	0.03400
40	1.48886	0.67165	48.88637	0.02046	32.83469	0.03046
45	1.56481	0.63905	56.48107	0.01771	36.09451	0.02771
50	1.64463	0.60804	64.46318	0.01551	39.19612	0.02551
55	1.72852	0.57853	72.85246	0.01373	42.14719	0.02373
60	1.81670	0.55045	81.66967	0.01224	44.95504	0.02224
65	1.90937	0.52373	90.93665	0.01100	47.62661	0.02100
70	2.00676	0.49831	100.67634	0.00993	50.16851	0.01993
75	2.10913	0.47413	110.91285	0.00902	52.58705	0.01902
80	2.21672	0.45112	121.67152	0.00822	54.88821	0.01822
85	2.32979	0.42922	132.97900	0.00752	57.07768	0.01752
90	2.44863	0.40839	144.86327	0.00690	59.16088	0.01690
95	2.57354	0.38857	157.35376	0.00636	61.14298	0.01636
100	2.70481	0.36971	170.48138	0.00587	63.02888	0.01587

n	$(1+i)^n$	$(1+i)^{-n}$	$s_{\bar{n}}$	R'_n	$_nt_0$	R''_n
1	1.01250	0.98765	1.00000	1.00000	0.98765	1.01250
2	1.02516	0.97546	2.01250	0.49689	1.96312	0.50939
3	1.03797	0.96342	3.03766	0.32920	2.92653	0.34170
4	1.05095	0.95152	4.07563	0.24536	3.87806	0.25786
5	1.06408	0.93978	5.12657	0.19506	4.81784	0.20756
6	1.07738	0.92817	6.19065	0.16153	5.74601	0.17403
7	1.09085	0.91672	7.26804	0.13759	6.66273	0.15009
8	1.10449	0.90540	8.35889	0.11963	7.56812	0.13213
9	1.11829	0.89422	9.46337	0.10567	8.46234	0.11817
10	1.13227	0.88318	10.58167	0.09450	9.34553	0.10700
11	1.14642	0.87228	11.71394	0.08537	10.21780	0.09787
12	1.16075	0.86151	12.86036	0.07776	11.07931	0.09026
13	1.17526	0.85087	14.02112	0.07132	11.93018	0.08382
14	1.18995	0.84037	15.19638	0.06581	12.77055	0.07831
15	1.20483	0.82999	16.38633	0.06103	13.60055	0.07353
16	1.21989	0.81975	17.59116	0.05685	14.42029	0.06935
17	1.23514	0.80963	18.81105	0.05316	15.22992	0.06566
18	1.25058	0.79963	20.04619	0.04988	16.02955	0.06238
19	1.26621	0.78976	21.29677	0.04696	16.81931	0.05946
20	1.28204	0.78001	22.56298	0.04432	17.59932	0.05682
21	1.29806	0.77038	23.84502	0.04194	18.36969	0.05444
22	1.31429	0.76087	25.14308	0.03977	19.13056	0.05227
23	1.33072	0.75147	26.45737	0.03780	19.88204	0.05030
24	1.34735	0.74220	27.78808	0.03599	20.62423	0.04849
25	1.36419	0.73303	29.13544	0.03432	21.35727	0.04682
26	1.38125	0.72398	30.49963	0.03279	22.08125	0.04529
27	1.39851	0.71505	31.88087	0.03137	22.79630	0.04387
28	1.41599	0.70622	33.27938	0.03005	23.50252	0.04255
29	1.43369	0.69750	34.69538	0.02882	24.20002	0.04132
30	1.45161	0.68889	36.12907	0.02768	24.88891	0.04018
35	1.54464	0.64740	43.57087	0.02295	28.20786	0.03545
40	1.64362	0.60841	51.48956	0.01942	31.32693	0.03192
45	1.74895	0.57177	59.91569	0.01669	34.25817	0.02919
50	1.86102	0.53734	68.88179	0.01452	37.01288	0.02702
55	1.98028	0.50498	78.42246	0.01275	39.60169	0.02525
60	2.10718	0.47457	88.57451	0.01129	42.03459	0.02379
65	2.24221	0.44599	99.37713	0.01006	44.32098	0.02256
70	2.38590	0.41913	110.87200	0.00902	46.46968	0.02152
75	2.53879	0.39389	123.10349	0.00812	48.48897	0.02062
80	2.70148	0.37017	136.11880	0.00735	50.38666	0.01985
85	2.87460	0.34787	149.96815	0.00667	52.17006	0.01917
90	3.05881	0.32692	164.70501	0.00607	53.84606	0.01857
95	3.25483	0.30724	180.38623	0.00554	55.42113	0.01804
100	3.46340	0.28873	197.07234	0.00507	56.90134	0.01757

n	$(1+i)^n$	$(1+i)^{-n}$	$s_{\bar{n}}$	R'_n	$_n l_0$	R''_n
1	1.01500	0.98522	1.00000	1.00000	0.98522	1.01500
2	1.03023	0.97066	2.01500	0.49628	1.95588	0.51128
3	1.04568	0.95632	3.04523	0.32838	2.91220	0.34338
4	1.06136	0.94218	4.09090	0.24444	3.85438	0.25944
5	1.07728	0.92826	5.15227	0.19409	4.78264	0.20909
6	1.09344	0.91454	6.22955	0.16053	5.69719	0.17553
7	1.10984	0.90103	7.32299	0.13656	6.59821	0.15156
8	1.12649	0.88771	8.43284	0.11858	7.48593	0.13358
9	1.14339	0.87459	9.55933	0.10461	8.36052	0.11961
10	1.16054	0.86167	10.70272	0.09343	9.22218	0.10843
11	1.17795	0.84893	11.86326	0.08429	10.07112	0.09929
12	1.19562	0.83639	13.04121	0.07668	10.90751	0.09168
13	1.21355	0.82403	14.23683	0.07024	11.73153	0.08524
14	1.23176	0.81185	15.45038	0.06472	12.54338	0.07972
15	1.25023	0.79985	16.68214	0.05994	13.34323	0.07494
16	1.26899	0.78803	17.93237	0.05577	14.13126	0.07077
17	1.28802	0.77639	19.20136	0.05208	14.90765	0.06708
18	1.30734	0.76491	20.48938	0.04881	15.67256	0.06381
19	1.32695	0.75361	21.79672	0.04588	16.42617	0.06088
20	1.34686	0.74247	23.12367	0.04325	17.16864	0.05825
21	1.36706	0.73150	24.47052	0.04087	17.90014	0.05587
22	1.38756	0.72069	25.83758	0.03870	18.62082	0.05370
23	1.40838	0.71004	27.22514	0.03673	19.33086	0.05173
24	1.42950	0.69954	28.63352	0.03492	20.03041	0.04992
25	1.45095	0.68921	30.06302	0.03326	20.71961	0.04826
26	1.47271	0.67902	31.51397	0.03173	21.39863	0.04673
27	1.49480	0.66899	32.98668	0.03032	22.06762	0.04532
28	1.51722	0.65910	34.48148	0.02900	22.72672	0.04400
29	1.53998	0.64936	35.99870	0.02778	23.37608	0.04278
30	1.56308	0.63976	37.53868	0.02664	24.01584	0.04164
35	1.68388	0.59387	45.59209	0.02193	27.07559	0.03693
40	1.81402	0.55126	54.26789	0.01843	29.91585	0.03343
45	1.95421	0.51171	63.61420	0.01572	32.55234	0.03072
50	2.10524	0.47500	73.68283	0.01357	34.99969	0.02857
55	2.26794	0.44093	84.52960	0.01183	37.27147	0.02683
60	2.44322	0.40930	96.21465	0.01039	39.38027	0.02539
65	2.63204	0.37993	108.80277	0.00919	41.33779	0.02419
70	2.83546	0.35268	122.36375	0.00817	43.15487	0.02317
75	3.05459	0.32738	136.97278	0.00730	44.84160	0.02230
80	3.29066	0.30389	152.71085	0.00655	46.40732	0.02155
85	3.54498	0.28209	169.66523	0.00589	47.86072	0.02089
90	3.81895	0.26185	187.92990	0.00532	49.20985	0.02032
95	4.11409	0.24307	207.60614	0.00482	50.46220	0.01982
100	4.43205	0.22563	228.80304	0.00437	51.62470	0.01937

n	$(1+i)^n$	$(1+i)^{-n}$	$s_{\bar{n}}$	R'_n	$_n t_0$	R''_n
1	1.01750	0.98280	1.00000	1.00000	0.98280	1.01750
2	1.03531	0.96590	2.01750	0.49566	1.94870	0.51316
3	1.05342	0.94929	3.05281	0.32757	2.89798	0.34507
4	1.07186	0.93296	4.10623	0.24353	3.83094	0.26103
5	1.09062	0.91691	5.17809	0.19312	4.74786	0.21062
6	1.10970	0.90114	6.26871	0.15952	5.64900	0.17702
7	1.12912	0.88564	7.37841	0.13553	6.53464	0.15303
8	1.14888	0.87041	8.50753	0.11754	7.40505	0.13504
9	1.16899	0.85544	9.65641	0.10356	8.26049	0.12106
10	1.18944	0.84073	10.82540	0.09238	9.10122	0.10988
11	1.21026	0.82627	12.01484	0.08323	9.92749	0.10073
12	1.23144	0.81206	13.22510	0.07561	10.73955	0.09311
13	1.25299	0.79809	14.45654	0.06917	11.53764	0.08667
14	1.27492	0.78436	15.70953	0.06366	12.32201	0.08116
15	1.29723	0.77087	16.98445	0.05888	13.09288	0.07638
16	1.31993	0.75762	18.28168	0.05470	13.85050	0.07220
17	1.34303	0.74459	19.60161	0.05102	14.59508	0.06852
18	1.36653	0.73178	20.94463	0.04774	15.32686	0.06524
19	1.39045	0.71919	22.31117	0.04482	16.04606	0.06232
20	1.41478	0.70682	23.70161	0.04219	16.75288	0.05969
21	1.43954	0.69467	25.11639	0.03981	17.44755	0.05731
22	1.46473	0.68272	26.55593	0.03766	18.13027	0.05516
23	1.49036	0.67098	28.02065	0.04069	18.80125	0.05819
24	1.51644	0.65944	29.51102	0.03389	19.46069	0.05139
25	1.54298	0.64810	31.02746	0.03223	20.10878	0.04973
26	1.56998	0.63695	32.57044	0.03070	20.74573	0.04820
27	1.59746	0.62599	34.14042	0.02929	21.37173	0.04679
28	1.62541	0.61523	35.73788	0.02798	21.98695	0.04548
29	1.65386	0.60465	37.36329	0.02676	22.59160	0.04426
30	1.68280	0.59425	39.01715	0.02563	23.18585	0.04313
35	1.83529	0.54487	47.73084	0.02095	26.00725	0.03845
40	2.00160	0.49960	57.23413	0.01747	28.59423	0.03497
45	2.18298	0.45809	67.59858	0.01479	30.96626	0.03229
50	2.38079	0.42003	78.90222	0.01267	33.14121	0.03017
55	2.59653	0.38513	91.23016	0.01096	35.13545	0.02846
60	2.83182	0.35313	104.67522	0.00955	36.96399	0.02705
65	3.08843	0.32379	119.33861	0.00838	38.64060	0.02588
70	3.36829	0.29689	135.33076	0.00739	40.17790	0.02489
75	3.67351	0.27222	152.77206	0.00655	41.58748	0.02405
80	4.00639	0.24960	171.79382	0.00582	42.87993	0.02332
85	4.36944	0.22886	192.53928	0.00519	44.06500	0.02269
90	4.76538	0.20985	215.16462	0.00465	45.15161	0.02215
95	5.19720	0.19241	239.84018	0.00417	46.14793	0.02167
100	5.66816	0.17642	266.75177	0.00375	47.06147	0.02125

n	$(1 + i)^n$	$(1 + i)^{-n}$	$s_{\bar{n}}$	R'_n	nl_0	R''_n
1	1.02000	0.98039	1.00000	1.00000	0.98039	1.02000
2	1.04040	0.96117	2.02000	0.49505	1.94156	0.51505
3	1.06121	0.94232	3.06040	0.32675	2.88388	0.34675
4	1.08243	0.92385	4.12161	0.24262	3.80773	0.26262
5	1.10408	0.90573	5.20404	0.19216	4.71346	0.21216
6	1.12616	0.88797	6.30812	0.15853	5.60143	0.17853
7	1.14869	0.87056	7.43428	0.13451	6.47199	0.15451
8	1.17166	0.85349	8.58297	0.11651	7.32548	0.13651
9	1.19509	0.83676	9.75463	0.10252	8.16224	0.12252
10	1.21899	0.82035	10.94972	0.09133	8.98259	0.11133
11	1.24337	0.80426	12.16872	0.08218	9.78685	0.10218
12	1.26824	0.78849	13.41209	0.07456	10.57534	0.09456
13	1.29361	0.77303	14.68033	0.06812	11.34837	0.08812
14	1.31948	0.75788	15.97394	0.06260	12.10625	0.08260
15	1.34587	0.74301	17.29342	0.05783	12.84926	0.07783
16	1.37279	0.72845	18.63929	0.05365	13.57771	0.07365
17	1.40024	0.71416	20.01207	0.04997	14.29187	0.06997
18	1.42825	0.70016	21.41231	0.04670	14.99203	0.06670
19	1.45681	0.68643	22.84056	0.04378	15.67846	0.06378
20	1.48595	0.67297	24.29737	0.04116	16.35143	0.06116
21	1.51567	0.65978	25.78332	0.03878	17.01121	0.05878
22	1.54598	0.64684	27.29898	0.03663	17.65805	0.05663
23	1.57690	0.63416	28.84496	0.03467	18.29220	0.05467
24	1.60844	0.62172	30.42186	0.03287	18.91393	0.05287
25	1.64061	0.60953	32.03030	0.03122	19.52346	0.05122
26	1.67342	0.59758	33.67091	0.02970	20.12104	0.04970
27	1.70689	0.58586	35.34432	0.02829	20.70690	0.04829
28	1.74102	0.57437	37.05121	0.02699	21.28127	0.04699
29	1.77584	0.56311	38.79223	0.02578	21.84438	0.04578
30	1.81136	0.55207	40.56808	0.02465	22.39646	0.04465
35	1.99989	0.50003	49.99448	0.02000	24.99862	0.04000
40	2.20804	0.45289	60.40198	0.01656	27.35548	0.03656
45	2.43785	0.41020	71.89271	0.01391	29.49016	0.03391
50	2.69159	0.37153	84.57940	0.01182	31.42361	0.03182
55	2.97173	0.33650	98.58653	0.01014	33.17479	0.03014
60	3.28103	0.30478	114.05154	0.00877	34.76089	0.02877
65	3.62252	0.27605	131.12616	0.00763	36.19747	0.02763
70	3.99956	0.25003	149.97791	0.00667	37.49862	0.02667
75	4.41584	0.22646	170.79177	0.00586	38.67711	0.02586
80	4.87544	0.20511	193.77196	0.00516	39.74451	0.02516
85	5.38288	0.18577	219.14394	0.00456	40.71129	0.02456
90	5.94313	0.16826	247.15666	0.00405	41.58693	0.02405
95	6.56170	0.15240	278.08496	0.00360	42.38002	0.02360
100	7.24465	0.13803	312.23231	0.00320	43.09835	0.02320

n	$(1+i)^n$	$(1+i)^{-n}$	$s_{\bar{n}}$	R'_n	$_n t_0$	R''_n
1	1.02500	0.97561	1.00000	1.00000	0.97561	1.02500
2	1.05063	0.95181	2.02500	0.49383	1.92742	0.51883
3	1.07689	0.92860	3.07563	0.32514	2.85602	0.35014
4	1.10381	0.90595	4.15252	0.24082	3.76197	0.26582
5	1.13141	0.88385	5.25633	0.19025	4.64583	0.21525
6	1.15969	0.86230	6.38774	0.15655	5.50813	0.18155
7	1.18869	0.84127	7.54743	0.13250	6.34939	0.15750
8	1.21840	0.82075	8.73612	0.11447	7.17014	0.13947
9	1.24886	0.80073	9.95452	0.10046	7.97087	0.12546
10	1.28008	0.78120	11.20338	0.08926	8.75206	0.11426
11	1.31209	0.76214	12.48347	0.08011	9.51421	0.10511
12	1.34489	0.74356	13.79555	0.07249	10.25776	0.09749
13	1.37851	0.72542	15.14044	0.06605	10.98318	0.09105
14	1.41297	0.70773	16.51895	0.06054	11.69091	0.08554
15	1.44830	0.69047	17.93193	0.05577	12.38138	0.08077
16	1.48451	0.67362	19.38022	0.05160	13.05500	0.07660
17	1.52162	0.65720	20.86473	0.04793	13.71220	0.07293
18	1.55966	0.64117	22.38635	0.04467	14.35336	0.06967
19	1.59865	0.62553	23.94601	0.04176	14.97889	0.06676
20	1.63862	0.61027	25.54466	0.03915	15.58916	0.06415
21	1.67958	0.59539	27.18327	0.03679	16.18455	0.06179
22	1.72157	0.58086	28.86286	0.03465	16.76541	0.05965
23	1.76461	0.56670	30.58443	0.03270	17.33211	0.05770
24	1.80873	0.55288	32.34904	0.03091	17.88499	0.05591
25	1.85394	0.53939	34.15776	0.02928	18.42438	0.05428
26	1.90029	0.52623	36.01171	0.02777	18.95061	0.05277
27	1.94780	0.51340	37.91200	0.02638	19.46401	0.05138
28	1.99650	0.50088	39.85980	0.02509	19.96489	0.05009
29	2.04641	0.48866	41.85630	0.02389	20.45355	0.04889
30	2.09757	0.47674	43.90270	0.02278	20.93029	0.04778
35	2.37321	0.42137	54.92821	0.01821	23.14516	0.04321
40	2.68506	0.37243	67.40255	0.01484	25.10278	0.03984
45	3.03790	0.32917	81.51613	0.01227	26.83302	0.03727
50	3.43711	0.29094	97.48435	0.01026	28.36231	0.03526
55	3.88877	0.25715	115.55092	0.00865	29.71398	0.03365
60	4.39979	0.22728	135.99159	0.00735	30.90866	0.03235
65	4.97796	0.20089	159.11833	0.00628	31.96458	0.03128
70	5.63210	0.17755	185.28411	0.00540	32.89786	0.03040
75	6.37221	0.15693	214.88830	0.00465	33.72274	0.02965
80	7.20957	0.13870	248.38271	0.00403	34.45182	0.02903
85	8.15696	0.12259	286.27857	0.00349	35.09621	0.02849
90	9.22886	0.10836	329.15425	0.00304	35.66577	0.02804
95	10.44160	0.09577	377.66415	0.00265	36.16917	0.02765
100	11.81372	0.08465	432.54865	0.00231	36.61411	0.02731

n	$(1+i)^n$	$(1+i)^{-n}$	$s_{\bar{n}}$	R'_n	$_n t_0$	R''_n
1	1.03000	0.97087	1.00000	1.00000	0.97087	1.03000
2	1.06090	0.94260	2.03000	0.49261	1.91347	0.52261
3	1.09273	0.91514	3.09090	0.32353	2.82861	0.35353
4	1.12551	0.88849	4.18363	0.23903	3.71710	0.26903
5	1.15927	0.86261	5.30914	0.18835	4.57971	0.21835
6	1.19405	0.83748	6.46841	0.15460	5.41719	0.18460
7	1.22987	0.81309	7.66246	0.13051	6.23028	0.16051
8	1.26677	0.78941	8.89234	0.11246	7.01969	0.14246
9	1.30477	0.76642	10.15911	0.09843	7.78611	0.12843
10	1.34392	0.74409	11.46388	0.08723	8.53020	0.11723
11	1.38423	0.72242	12.80780	0.07808	9.25262	0.10808
12	1.42576	0.70138	14.19203	0.07046	9.95400	0.10046
13	1.46853	0.68095	15.61779	0.06403	10.63496	0.09403
14	1.51259	0.66112	17.08632	0.05853	11.29607	0.08853
15	1.55797	0.64186	18.59891	0.05377	11.93794	0.08377
16	1.60471	0.62317	20.15688	0.04961	12.56110	0.07961
17	1.65285	0.60502	21.76159	0.04595	13.16612	0.07595
18	1.70243	0.58739	23.41444	0.04271	13.75351	0.07271
19	1.75351	0.57029	25.11687	0.03981	14.32380	0.06981
20	1.80611	0.55368	26.87037	0.03722	14.87747	0.06722
21	1.86029	0.53755	28.67649	0.03487	15.41502	0.06487
22	1.91610	0.52189	30.53678	0.03275	15.93692	0.06275
23	1.97359	0.50669	32.45288	0.03081	16.44361	0.06081
24	2.03279	0.49193	34.42647	0.02905	16.93554	0.05905
25	2.09378	0.47761	36.45926	0.02743	17.41315	0.05743
26	2.15659	0.46369	38.55304	0.02594	17.87684	0.05594
27	2.22129	0.45019	40.70963	0.02456	18.32703	0.05456
28	2.28793	0.43708	42.93092	0.02329	18.76411	0.05329
29	2.35657	0.42435	45.21885	0.02211	19.18845	0.05211
30	2.42726	0.41199	47.57542	0.02102	19.60044	0.05102
35	2.81386	0.35538	60.46208	0.01654	21.48722	0.04654
40	3.26204	0.30656	75.40126	0.01326	23.11477	0.04326
45	3.78160	0.26444	92.71986	0.01079	24.51871	0.04079
50	4.38391	0.22811	112.79687	0.00887	25.72976	0.03887
55	5.08215	0.19677	136.07162	0.00735	26.77443	0.03735
60	5.89160	0.16973	163.05344	0.00613	27.67556	0.03613
65	6.82998	0.14641	194.33276	0.00515	28.45289	0.03515
70	7.91782	0.12630	230.59406	0.00434	29.12342	0.03434
75	9.17893	0.10895	272.63086	0.00367	29.70183	0.03367
80	10.64089	0.09398	321.36302	0.00311	30.20076	0.03311
85	12.33571	0.08107	377.85695	0.00265	30.63115	0.03265
90	14.30047	0.06993	443.34890	0.00226	31.00241	0.03226
95	16.57816	0.06032	519.27203	0.00193	31.32266	0.03193
100	19.21863	0.05203	607.28773	0.00165	31.59891	0.03165

n	$(1+i)^n$	$(1+i)^{-n}$	$s_{\bar{n}}$	R'_n	$_n t_0$	R''_n
1	1.03500	0.96618	1.00000	1.00000	0.96618	1.03500
2	1.07123	0.93351	2.03500	0.49140	1.89969	0.52640
3	1.10872	0.90194	3.10623	0.32193	2.80164	0.35693
4	1.14752	0.87144	4.21494	0.23725	3.67308	0.27225
5	1.18769	0.84197	5.36247	0.18648	4.51505	0.22148
6	1.22926	0.81350	6.55015	0.15267	5.32855	0.18767
7	1.27228	0.78599	7.77941	0.12854	6.11454	0.16354
8	1.31681	0.75941	9.05169	0.11048	6.87396	0.14548
9	1.36290	0.73373	10.36850	0.09645	7.60769	0.13145
10	1.41060	0.70892	11.73139	0.08524	8.31661	0.12024
11	1.45997	0.68495	13.14199	0.07609	9.00155	0.11109
12	1.51107	0.66178	14.60196	0.06848	9.66333	0.10348
13	1.56396	0.63940	16.11303	0.06206	10.30274	0.09706
14	1.61869	0.61778	17.67699	0.05657	10.92052	0.09157
15	1.67535	0.59689	19.29568	0.05183	11.51741	0.08683
16	1.73399	0.57671	20.97103	0.04768	12.09412	0.08268
17	1.79468	0.55720	22.70502	0.04404	12.65132	0.07904
18	1.85749	0.53836	24.49969	0.04082	13.18968	0.07582
19	1.92250	0.52016	26.35718	0.03794	13.70984	0.07294
20	1.98979	0.50257	28.27968	0.03536	14.21240	0.07036
21	2.05943	0.48557	30.26947	0.03304	14.69797	0.06804
22	2.13151	0.46915	32.32890	0.03093	15.16712	0.06593
23	2.20611	0.45329	34.46041	0.02902	15.62041	0.06402
24	2.28333	0.43796	36.66653	0.02727	16.05837	0.06227
25	2.36324	0.42315	38.94986	0.02567	16.48151	0.06067
26	2.44596	0.40884	41.31310	0.02421	16.89035	0.05921
27	2.53157	0.39501	43.75906	0.02285	17.28536	0.05785
28	2.62017	0.38165	46.29063	0.02160	17.66702	0.05660
29	2.71188	0.36875	48.91080	0.02045	18.03577	0.05545
30	2.80679	0.35628	51.62268	0.01937	18.39205	0.05437
35	3.33359	0.29998	66.67401	0.01500	20.00066	0.05000
40	3.95926	0.25257	84.55028	0.01183	21.35507	0.04683
45	4.70236	0.21266	105.78167	0.00945	22.49545	0.04445
50	5.58492	0.17905	130.99791	0.00763	23.45562	0.04263
55	6.63314	0.15076	160.94689	0.00621	24.26405	0.04121
60	7.87809	0.12693	196.51688	0.00509	24.94473	0.04009
65	9.35670	0.10688	238.76288	0.00419	25.51785	0.03919
70	11.11283	0.08999	288.93786	0.00346	26.00040	0.03846
75	13.19855	0.07577	348.53001	0.00287	26.40669	0.03787
80	15.67574	0.06379	419.30679	0.00238	26.74878	0.03738
85	18.61786	0.05371	503.36739	0.00199	27.03680	0.03699
90	22.11218	0.04522	603.20503	0.00166	27.27932	0.03666
95	26.26233	0.03808	721.78082	0.00139	27.48350	0.03639
100	31.19141	0.03206	862.61166	0.00116	27.65543	0.03616

n	$(1 + i)^n$	$(1 + i)^{-n}$	$s_{\bar{n}}$	R'_n	$_nl_0$	R''_n
1	1.04000	0.96154	1.00000	1.00000	0.96154	1.04000
2	1.08160	0.92456	2.04000	0.49020	1.88609	0.53020
3	1.12486	0.88900	3.12160	0.32035	2.77509	0.36035
4	1.16986	0.85480	4.24646	0.23549	3.62990	0.27549
5	1.21665	0.82193	5.41632	0.18463	4.45182	0.22463
6	1.26532	0.79031	6.63298	0.15076	5.24214	0.19076
7	1.31593	0.75992	7.89829	0.12661	6.00206	0.16661
8	1.36857	0.73069	9.21423	0.10853	6.73274	0.14853
9	1.42331	0.70259	10.58280	0.09449	7.43533	0.13449
10	1.48024	0.67556	12.00611	0.08329	8.11090	0.12329
11	1.53945	0.64958	13.48635	0.07415	8.76048	0.11415
12	1.60103	0.62460	15.02581	0.06655	9.38507	0.10655
13	1.66507	0.60057	16.62684	0.06014	9.98565	0.10014
14	1.73168	0.57748	18.29191	0.05467	10.56312	0.09467
15	1.80094	0.55526	20.02359	0.04994	11.11839	0.08994
16	1.87298	0.53391	21.82453	0.04582	11.65230	0.08582
17	1.94790	0.51337	23.69751	0.04220	12.16567	0.08220
18	2.02582	0.49363	25.64541	0.03899	12.65930	0.07899
19	2.10685	0.47464	27.67123	0.03614	13.13394	0.07614
20	2.19112	0.45639	29.77808	0.03358	13.59033	0.07358
21	2.27877	0.43883	31.96920	0.03128	14.02916	0.07128
22	2.36992	0.42196	34.24797	0.02920	14.45112	0.06920
23	2.46472	0.40573	36.61789	0.02731	14.85684	0.06731
24	2.56330	0.39012	39.08260	0.02559	15.24696	0.06559
25	2.66584	0.37512	41.64591	0.02401	15.62208	0.06401
26	2.77247	0.36069	44.31174	0.02257	15.98277	0.06257
27	2.88337	0.34682	47.08421	0.02124	16.32959	0.06124
28	2.99870	0.33348	49.96758	0.02001	16.66306	0.06001
29	3.11865	0.32065	52.96629	0.01888	16.98371	0.05888
30	3.24340	0.30832	56.08494	0.01783	17.29203	0.05783
35	3.94609	0.25342	73.65222	0.01358	18.66461	0.05358
40	4.80102	0.20829	95.02552	0.01052	19.79277	0.05052
45	5.84118	0.17120	121.02939	0.00826	20.72004	0.04826
50	7.10668	0.14071	152.66708	0.00655	21.48218	0.04655
55	8.64637	0.11566	191.15917	0.00523	22.10861	0.04523
60	10.51963	0.09506	237.99069	0.00420	22.62349	0.04420
65	12.79874	0.07813	294.96838	0.00339	23.04668	0.04339
70	15.57162	0.06422	364.29046	0.00275	23.39451	0.04275
75	18.94525	0.05278	448.63137	0.00223	23.68041	0.04223
80	23.04980	0.04338	551.24498	0.00181	23.91539	0.04181
85	28.04360	0.03566	676.09012	0.00148	24.10853	0.04148
90	34.11933	0.02931	827.98333	0.00121	24.26728	0.04121
95	41.51139	0.02409	1012.78465	0.00099	24.39776	0.04099
100	50.50495	0.01980	1237.62370	0.00081	24.50500	0.04081

n	$(1+i)^n$	$(1+i)^{-n}$	$s_{\overline{n}}$	R'_n	$_nt_0$	R''_n
1	1.04500	0.95694	1.00000	1.00000	0.95694	1.04500
2	1.09203	0.91573	2.04500	0.48900	1.87267	0.53400
3	1.14117	0.87630	3.13703	0.31877	2.74896	0.36377
4	1.19252	0.83856	4.27819	0.23374	3.58753	0.27874
5	1.24618	0.80245	5.47071	0.18279	4.38998	0.22779
6	1.30226	0.76790	6.71689	0.14888	5.15787	0.19388
7	1.36086	0.73483	8.01915	0.12470	5.89270	0.16970
8	1.42210	0.70319	9.38001	0.10661	6.59589	0.15161
9	1.48610	0.67290	10.80211	0.09257	7.26879	0.13757
10	1.55297	0.64393	12.28821	0.08138	7.91272	0.12638
11	1.62285	0.61620	13.84118	0.07225	8.52892	0.11725
12	1.69588	0.58966	15.46403	0.06467	9.11858	0.10967
13	1.77220	0.56427	17.15991	0.05828	9.68285	0.10328
14	1.85194	0.53997	18.93211	0.05282	10.22283	0.09782
15	1.93528	0.51672	20.78405	0.04811	10.73955	0.09311
16	2.02237	0.49447	22.71934	0.04402	11.23402	0.08902
17	2.11338	0.47318	24.74171	0.04042	11.70719	0.08542
18	2.20848	0.45280	26.85508	0.03724	12.15999	0.08224
19	2.30786	0.43330	29.06356	0.03441	12.59329	0.07941
20	2.41171	0.41464	31.37142	0.03188	13.00794	0.07688
21	2.52024	0.39679	33.78314	0.02960	13.40472	0.07460
22	2.63365	0.37970	36.30338	0.02755	13.78442	0.07255
23	2.75217	0.36335	38.93703	0.02568	14.14777	0.07068
24	2.87601	0.34770	41.68920	0.02399	14.49548	0.06899
25	3.00543	0.33273	44.56521	0.02244	14.82821	0.06744
26	3.14068	0.31840	47.57064	0.02102	15.14661	0.06602
27	3.28201	0.30469	50.71132	0.01972	15.45130	0.06472
28	3.42970	0.29157	53.99333	0.01852	15.74287	0.06352
29	3.58404	0.27902	57.42303	0.01741	16.02189	0.06241
30	3.74532	0.26700	61.00707	0.01639	16.28889	0.06139
35	4.66735	0.21425	81.49662	0.01227	17.46101	0.05727
40	5.81636	0.17193	107.03032	0.00934	18.40158	0.05434
45	7.24825	0.13796	138.84997	0.00720	19.15635	0.05220
50	9.03264	0.11071	178.50303	0.00560	19.76201	0.05060
55	11.25631	0.08884	227.91796	0.00439	20.24802	0.04939
60	14.02741	0.07129	289.49795	0.00345	20.63802	0.04845
65	17.48070	0.05721	366.23783	0.00273	20.95098	0.04773
70	21.78414	0.04590	461.86968	0.00217	21.20211	0.04717
75	27.14700	0.03684	581.04436	0.00172	21.40363	0.04672
80	33.83010	0.02956	729.55770	0.00137	21.56534	0.04637
85	42.15846	0.02372	914.63234	0.00109	21.69511	0.04609
90	52.53711	0.01903	1145.26901	0.00087	21.79924	0.04587
95	65.47079	0.01527	1432.68426	0.00070	21.88280	0.04570
100	81.58852	0.01226	1790.85596	0.00056	21.94985	0.04556

n	$(1+i)^n$	$(1+i)^{-n}$	$s_{\bar{n}}$	R'_n	$_nl_0$	R''_n
1	1.05000	0.95238	1.00000	1.00000	0.95238	1.05000
2	1.10250	0.90703	2.05000	0.48780	1.85941	0.53780
3	1.15763	0.86384	3.15250	0.31721	2.72325	0.36721
4	1.21551	0.82270	4.31013	0.23201	3.54595	0.28201
5	1.27628	0.78353	5.52563	0.18097	4.32948	0.23097
6	1.34010	0.74622	6.80191	0.14702	5.07569	0.19702
7	1.40710	0.71068	8.14201	0.12282	5.78637	0.17282
8	1.47746	0.67684	9.54911	0.10472	6.46321	0.15472
9	1.55133	0.64461	11.02656	0.09069	7.10782	0.14069
10	1.62889	0.61391	12.57789	0.07950	7.72173	0.12950
11	1.71034	0.58468	14.20679	0.07039	8.30641	0.12039
12	1.79586	0.55684	15.91713	0.06283	8.86325	0.11283
13	1.88565	0.53032	17.71298	0.05646	9.39357	0.10646
14	1.97993	0.50507	19.59863	0.05102	9.89864	0.10102
15	2.07893	0.48102	21.57856	0.04634	10.37966	0.09634
16	2.18287	0.45811	23.65749	0.04227	10.83777	0.09227
17	2.29202	0.43630	25.84037	0.03870	11.27407	0.08870
18	2.40662	0.41552	28.13238	0.03555	11.68959	0.08555
19	2.52695	0.39573	30.53900	0.03275	12.08532	0.08275
20	2.65330	0.37689	33.06595	0.03024	12.46221	0.08024
21	2.78596	0.35894	35.71925	0.02800	12.82115	0.07800
22	2.92526	0.34185	38.50521	0.02597	13.16300	0.07597
23	3.07152	0.32557	41.43048	0.02414	13.48857	0.07414
24	3.22510	0.31007	44.50200	0.02247	13.79864	0.07247
25	3.38635	0.29530	47.72710	0.02095	14.09394	0.07095
26	3.55567	0.28124	51.11345	0.01956	14.37519	0.06956
27	3.73346	0.26785	54.66913	0.01829	14.64303	0.06829
28	3.92013	0.25509	58.40258	0.01712	14.89813	0.06712
29	4.11614	0.24295	62.32271	0.01605	15.14107	0.06605
30	4.32194	0.23138	66.43885	0.01505	15.37245	0.06505
35	5.51602	0.18129	90.32031	0.01107	16.37419	0.06107
40	7.03999	0.14205	120.79977	0.00828	17.15909	0.05828
45	8.98501	0.11130	159.70016	0.00626	17.77407	0.05626
50	11.46740	0.08720	209.34800	0.00478	18.25593	0.05478
55	14.63563	0.06833	272.71262	0.00367	18.63347	0.05367
60	18.67919	0.05354	353.58372	0.00283	18.92929	0.05283
65	23.83990	0.04195	456.79801	0.00219	19.16107	0.05219
70	30.42643	0.03287	588.52851	0.00170	19.34268	0.05170
75	38.83269	0.02575	756.65372	0.00132	19.48497	0.05132
80	49.56144	0.02018	971.22882	0.00103	19.59646	0.05103
85	63.25435	0.01581	1245.08707	0.00080	19.68382	0.05080
90	80.73037	0.01239	1594.60730	0.00063	19.75226	0.05063
95	103.03468	0.00971	2040.69353	0.00049	19.80589	0.05049
100	131.50126	0.00760	2610.02516	0.00038	19.84791	0.05038

n	$(1+i)^n$	$(1+i)^{-n}$	$s_{\bar{n}}$	R'_n	$_n t_0$	R''_n
1	1.05500	0.94787	1.00000	1.00000	0.94787	1.05500
2	1.11303	0.89845	2.05500	0.48662	1.84632	0.54162
3	1.17424	0.85161	3.16803	0.31565	2.69793	0.37065
4	1.23882	0.80722	4.34227	0.23029	3.50515	0.28529
5	1.30696	0.76513	5.58109	0.17918	4.27028	0.23418
6	1.37884	0.72525	6.88805	0.14518	4.99553	0.20018
7	1.45468	0.68744	8.26689	0.12096	5.68297	0.17596
8	1.53469	0.65160	9.72157	0.10286	6.33457	0.15786
9	1.61909	0.61763	11.25626	0.08884	6.95220	0.14384
10	1.70814	0.58543	12.87535	0.07767	7.53763	0.13267
11	1.80209	0.55491	14.58350	0.06857	8.09254	0.12357
12	1.90121	0.52598	16.38559	0.06103	8.61852	0.11603
13	2.00577	0.49856	18.28680	0.05468	9.11708	0.10968
14	2.11609	0.47257	20.29257	0.04928	9.58965	0.10428
15	2.23248	0.44793	22.40866	0.04463	10.03758	0.09963
16	2.35526	0.42458	24.64114	0.04058	10.46216	0.09558
17	2.48480	0.40245	26.99640	0.03704	10.86461	0.09204
18	2.62147	0.38147	29.48120	0.03392	11.24607	0.08892
19	2.76565	0.36158	32.10267	0.03115	11.60765	0.08615
20	2.91776	0.34273	34.86832	0.02868	11.95038	0.08368
21	3.07823	0.32486	37.78608	0.02646	12.27524	0.08146
22	3.24754	0.30793	40.86431	0.02447	12.58317	0.07947
23	3.42615	0.29187	44.11185	0.02267	12.87504	0.07767
24	3.61459	0.27666	47.53800	0.02104	13.15170	0.07604
25	3.81339	0.26223	51.15259	0.01955	13.41393	0.07455
26	4.02313	0.24856	54.96598	0.01819	13.66250	0.07319
27	4.24440	0.23560	58.98911	0.01695	13.89810	0.07195
28	4.47784	0.22332	63.23351	0.01581	14.12142	0.07081
29	4.72412	0.21168	67.71135	0.01477	14.33310	0.06977
30	4.98395	0.20064	72.43548	0.01381	14.53375	0.06881
35	6.51383	0.15352	100.25136	0.00997	15.39055	0.06497
40	8.51331	0.11746	136.60561	0.00732	16.04612	0.06232
45	11.12655	0.08988	184.11917	0.00543	16.54773	0.06043
50	14.54196	0.06877	246.21748	0.00406	16.93152	0.05906
55	19.00576	0.05262	327.37749	0.00305	17.22517	0.05805
60	24.83977	0.04026	433.45037	0.00231	17.44985	0.05731
65	32.46459	0.03080	572.08339	0.00175	17.62177	0.05675
70	42.42992	0.02357	753.27120	0.00133	17.75330	0.05633
75	55.45420	0.01803	990.07643	0.00101	17.85395	0.05601
80	72.47643	0.01380	1299.57139	0.00077	17.93095	0.05577
85	94.72379	0.01056	1704.06892	0.00059	17.98987	0.05559
90	123.80021	0.00808	2232.73102	0.00045	18.03495	0.05545
95	161.80192	0.00618	2923.67123	0.00034	18.06945	0.05534
100	211.46864	0.00473	3826.70247	0.00026	18.09584	0.05526

n	$(1+i)^n$	$(1+i)^{-n}$	$s_{\bar{n}}$	R'_n	$_nl_0$	R''_n
1	1.06000	0.94340	1.00000	1.00000	0.94340	1.06000
2	1.12360	0.89000	2.06000	0.48544	1.83339	0.54544
3	1.19102	0.83962	3.18360	0.31411	2.67301	0.37411
4	1.26248	0.79209	4.37462	0.22859	3.46511	0.28859
5	1.33823	0.74726	5.63709	0.17740	4.21236	0.23740
6	1.41852	0.70496	6.97532	0.14336	4.91732	0.20336
7	1.50363	0.66506	8.39384	0.11914	5.58238	0.17914
8	1.59385	0.62741	9.89747	0.10104	6.20979	0.16104
9	1.68948	0.59190	11.49132	0.08702	6.80169	0.14702
10	1.79085	0.55839	13.18079	0.07587	7.36009	0.13587
11	1.89830	0.52679	14.97164	0.06679	7.88687	0.12679
12	2.01220	0.49697	16.86994	0.05928	8.38384	0.11928
13	2.13293	0.46884	18.88214	0.05296	8.85268	0.11296
14	2.26090	0.44230	21.01507	0.04758	9.29498	0.10758
15	2.39656	0.41727	23.27597	0.04296	9.71225	0.10296
16	2.54035	0.39365	25.67253	0.03895	10.10590	0.09895
17	2.69277	0.37136	28.21288	0.03544	10.47726	0.09544
18	2.85434	0.35034	30.90565	0.03236	10.82760	0.09236
19	3.02560	0.33051	33.75999	0.02962	11.15812	0.08962
20	3.20714	0.31180	36.78559	0.02718	11.46992	0.08718
21	3.39956	0.29416	39.99273	0.02500	11.76408	0.08500
22	3.60354	0.27751	43.39229	0.02305	12.04158	0.08305
23	3.81975	0.26180	46.99583	0.02128	12.30338	0.08128
24	4.04893	0.24698	50.81558	0.01968	12.55036	0.07968
25	4.29187	0.23300	54.86451	0.01823	12.78336	0.07823
26	4.54938	0.21981	59.15638	0.01690	13.00317	0.07690
27	4.82235	0.20737	63.70577	0.01570	13.21053	0.07570
28	5.11169	0.19563	68.52811	0.01459	13.40616	0.07459
29	5.41839	0.18456	73.63980	0.01358	13.59072	0.07358
30	5.74349	0.17411	79.05819	0.01265	13.76483	0.07265
35	7.68609	0.13011	111.43478	0.00897	14.49825	0.06897
40	10.28572	0.09722	154.76197	0.00646	15.04630	0.06646
45	13.76461	0.07265	212.74351	0.00470	15.45583	0.06470
50	18.42015	0.05429	290.33590	0.00344	15.76186	0.06344
55	24.65032	0.04057	394.17203	0.00254	15.99054	0.06254
60	32.98769	0.03031	533.12818	0.00188	16.16143	0.06188
65	44.14497	0.02265	719.08286	0.00139	16.28912	0.06139
70	59.07593	0.01693	967.93217	0.00103	16.38454	0.06103
75	79.05692	0.01265	1300.94868	0.00077	16.45585	0.06077
80	105.79599	0.00945	1746.59989	0.00057	16.50913	0.06057
85	141.57890	0.00706	2342.98174	0.00043	16.54895	0.06043
90	189.46451	0.00528	3141.07519	0.00032	16.57870	0.06032
95	253.54625	0.00394	4209.10425	0.00024	16.60093	0.06024
100	339.30208	0.00295	5638.36806	0.00018	16.61755	0.06018

n	$(1+i)^n$	$(1+i)^{-n}$	$s_{\bar{n}}$	R'_n	$_n l_0$	R''_n
1	1.06500	0.93897	1.00000	1.00000	0.93897	1.06500
2	1.13423	0.88166	2.06500	0.48426	1.82063	0.54926
3	1.20795	0.82785	3.19923	0.31258	2.64848	0.37758
4	1.28647	0.77732	4.40717	0.22690	3.42580	0.29190
5	1.37009	0.72988	5.69364	0.17563	4.15568	0.24063
6	1.45914	0.68533	7.06373	0.14157	4.84101	0.20657
7	1.55399	0.64351	8.52287	0.11733	5.48452	0.18233
8	1.65500	0.60423	10.07686	0.09924	6.08875	0.16424
9	1.76257	0.56735	11.73185	0.08524	6.65610	0.15024
10	1.87714	0.53273	13.49442	0.07410	7.18883	0.13910
11	1.99915	0.50021	15.37156	0.06506	7.68904	0..3006
12	2.12910	0.46968	17.37071	0.05757	8.15873	0.12257
13	2.26749	0 44102	19.49981	0.05128	8.59974	0.11628
14	2.41487	0.41410	21.76730	0.04594	9.01384	0.11094
15	2.57184	0.38883	24.18217	0.04135	9.40267	0.10635
16	2.73901	0.36510	26.75401	0.03738	9.76776	0.10238
17	2.91705	0.34281	29.49302	0.03391	10.11058	0.09891
18	3.10665	0.32190	32.41007	0.03085	10.43247	0.09585
19	3.30859	0.30224	35.51672	0.02816	10.73471	0.09316
20	3.52365	0.28380	38.82531	0.02576	11.01851	0.09076
21	3.75268	0.26648	42.34895	0.02361	11.28498	0.08861
22	3.99661	0.25021	46.10164	0.02169	11.53520	0.08669
23	4.25639	0.23494	50.09824	0.01996	11.77014	0.08496
24	4.53305	0.22060	54.35463	0.01840	11.99074	0.08340
25	4.82770	0.20714	58.88768	0.01698	12.19788	0.08198
26	5.14150	0.19450	63.71538	0.01569	12.39237	0.08069
27	5.47570	0.18263	68.85688	0.01452	12.57500	0.07952
28	5.83162	0.17148	74.33257	0.01345	12.74648	0.07845
29	6.21067	0.16101	80.16419	0.01247	12.90749	0.07747
30	6.61437	0.15119	86.37486	0.01158	13.05868	0.07658
35	9.06225	0.11035	124.03469	0.00806	13.68696	0.07306
40	12.41607	0.08054	175.63192	0.00569	14.14553	0.07069
45	17.01110	0.05879	246.32459	0.00406	14.48023	0.06906
50	23.30668	0.04291	343.17967	0.00291	14.72452	0.06791

n	$(1+i)^n$	$(1+i)^{-n}$	$s_{\bar{n}}$	R'_n	$_n l_0$	R''_n
1	1.07000	0.93458	1.00000	1.00000	0.93458	1.07000
2	1.14490	0.87344	2.07000	0.48309	1.80802	0.55309
3	1.22504	0.81630	3.21490	0.31105	2.62432	0.38105
4	1.31080	0.76290	4.43994	0.22523	3.38721	0.29523
5	1.40255	0.71299	5.75074	0.17389	4.10020	0.24389
6	1.50073	0.66634	7.15329	0.13980	4.76654	0.20980
7	1.60578	0.62275	8.65402	0.11555	5.38929	0.18555
8	1.71819	0.58201	10.25980	0.09747	5.97130	0.16747
9	1.83846	0.54393	11.97799	0.08349	6.51523	0.15349
10	1.96715	0.50835	13.81645	0.07238	7.02358	0.14238
11	2.10485	0.47509	15.78360	0.06336	7.49867	0.13336
12	2.25219	0.44401	17.88845	0.05590	7.94269	0.12590
13	2.40985	0.41496	20.14064	0.04965	8.35765	0.11965
14	2.57853	0.38782	22.55049	0.04434	8.74547	0.11434
15	2.75903	0.36245	25.12902	0.03979	9.10791	0.10979
16	2.95216	0.33873	27.88805	0.03586	9.44665	0.10586
17	3.15882	0.31657	30.84022	0.03243	9.76322	0.10243
18	3.37993	0.29586	33.99903	0.02941	10.05909	0.09941
19	3.61653	0.27651	37.37896	0.02675	10.33560	0.09675
20	3.86968	0.25842	40.99549	0.02439	10.59401	0.09439
21	4.14056	0.24151	44.86518	0.02229	10.83553	0.09229
22	4.43040	0.22571	49.00574	0.02041	11.06124	0.09041
23	4.74053	0.21095	53.43614	0.01871	11.27219	0.08871
24	5.07237	0.19715	58.17667	0.01719	11.46933	0.08719
25	5.42743	0.18425	63.24904	0.01581	11.65358	0.08581
26	5.80735	0.17220	68.67647	0.01456	11.82578	0.08456
27	6.21387	0.16093	74.48382	0.01343	11.98671	0.08343
28	6.64884	0.15040	80.69769	0.01239	12.13711	0.08239
29	7.11426	0.14056	87.34653	0.01145	12.27767	0.08145
30	7.61226	0.13137	94.46079	0.01059	12.40904	0.08059
35	10.67658	0.09366	138.23688	0.00723	12.94767	0.07723
40	14.97446	0.06678	199.63511	0.00501	13.33171	0.07501
45	21.00245	0.04761	285.74931	0.00350	13.60552	0.07350
50	29.45703	0.03395	406.52893	0.00246	13.80075	0.07246

n	$(1+i)^n$	$(1+i)^{-n}$	$s_{\bar{n}}$	R'_n	$_nt_0$	R''_n
1	1.08000	0.92593	1.00000	1.00000	0.92593	1.08000
2	1.16640	0.85734	2.08000	0.48077	1.78326	0.56077
3	1.25971	0.79383	3.24640	0.30803	2.57710	0.38803
4	1.36049	0.73503	4.50611	0.22192	3.31213	0.30192
5	1.46933	0.68058	5.86660	0.17046	3.99271	0.25046
6	1.58687	0.63017	7.33593	0.13632	4.62288	0.21632
7	1.71382	0.58349	8.92280	0.11207	5.20637	0.19207
8	1.85093	0.54027	10.63663	0.09401	5.74664	0.17401
9	1.99900	0.50025	12.48756	0.08008	6.24689	0.16008
10	2.15893	0.46319	14.48656	0.06903	6.71008	0.14903
11	2.33164	0.42888	16.64549	0.06008	7.13896	0.14008
12	2.51817	0.39711	18.97713	0.05270	7.53608	0.13270
13	2.71962	0.36770	21.49530	0.04652	7.90378	0.12652
14	2.93719	0.34046	24.21492	0.04130	8.24424	0.12130
15	3.17217	0.31524	27.15211	0.03683	8.55948	0.11683
16	3.42594	0.29189	30.32428	0.03298	8.85137	0.11298
17	3.70002	0.27027	33.75023	0.02963	9.12164	0.10963
18	3.99602	0.25025	37.45024	0.02670	9.37189	0.10670
19	4.31570	0.23171	41.44626	0.02413	9.60360	0.10413
20	4.66096	0.21455	45.76196	0.02185	9.81815	0.10185
21	5.03383	0.19866	50.42292	0.01983	10.01680	0.09983
22	5.43654	0.18394	55.45676	0.01803	10.20074	0.09803
23	5.87146	0.17032	60.89330	0.01642	10.37106	0.09642
24	6.34118	0.15770	66.76476	0.01498	10.52876	0.09498
25	6.84848	0.14602	73.10594	0.01368	10.67478	0.09368
26	7.39635	0.13520	79.95442	0.01251	10.80998	0.09251
27	7.98806	0.12519	87.35077	0.01145	10.93516	0.09145
28	8.62711	0.11591	95.33883	0.01049	11.05108	0.09049
29	9.31727	0.10733	103.96594	0.00962	11.15841	0.08962
30	10.06266	0.09938	113.28321	0.00883	11.25778	0.08883
35	14.78534	0.06763	172.31680	0.00580	11.65457	0.08580
40	21.72452	0.04603	259.05652	0.00386	11.92461	0.08386
45	31.92045	0.03133	386.50562	0.00259	12.10840	0.08259
50	46.90161	0.02132	573.77016	0.00174	12.23348	0.08174

n	$(1+i)^n$	$(1+i)^{-n}$	$s_{\bar{n}}$	R'_n	$_nl_0$	R''_n
1	1.10000	0.90909	1.00000	1.00000	0.90909	1.10000
2	1.21000	0.82645	2.10000	0.47619	1.73553	0.57619
3	1.33100	0.75131	3.31000	0.30211	2.48685	0.40211
4	1.46410	0.68301	4.64100	0.21547	3.16987	0.31547
5	1.61051	0.62092	6.10510	0.16380	3.79079	0.26380
6	1.77156	0.56447	7.71561	0.12961	4.35526	0.22961
7	1.94872	0.51316	9.48717	0.10541	4.86842	0.20541
8	2.14359	0.46651	11.43589	0.08744	5.33493	0.18744
9	2.35795	0.42410	13.57948	0.07364	5.75902	0.17364
10	2.59374	0.38554	15.93742	0.06275	6.14457	0.16275
11	2.85312	0.35049	18.53117	0.05396	6.49506	0.15396
12	3.13843	0.31863	21.38428	0.04676	6.81369	0.14676
13	3.45227	0.28966	24.52271	0.04078	7.10336	0.14078
14	3.79750	0.26333	27.97498	0.03575	7.36669	0.13575
15	4.17725	0.23939	31.77248	0.03147	7.60608	0.13147
16	4.59497	0.21763	35.94973	0.02782	7.82371	0.12782
17	5.05447	0.19784	40.54470	0.02466	8.02155	0.12466
18	5.55992	0.17986	45.59917	0.02193	8.20141	0.12193
19	6.11591	0.16351	51.15909	0.01955	8.36492	0.11955
20	6.72750	0.14864	57.27500	0.01746	8.51356	0.11746
21	7.40025	0.13513	64.00250	0.01562	8.64869	0.11562
22	8.14027	0.12285	71.40275	0.01401	8.77154	0.11401
23	8.95430	0.11168	79.54302	0.01257	8.88322	0.11257
24	9.84973	0.10153	88.49733	0.01130	8.98474	0.11130
25	10.83471	0.09230	98.34706	0.01017	9.07704	0.11017
26	11.91818	0.08391	109.18177	0.00916	9.16095	0.10916
27	13.10999	0.07628	121.09994	0.00826	9.23722	0.10826
28	14.42099	0.06934	134.20994	0.00745	9.30657	0.10745
29	15.86309	0.06304	148.63093	0.00673	9.36961	0.10673
30	17.44940	0.05731	164.49402	0.00608	9.42691	0.10608
35	28.10244	0.03558	271.02437	0.00369	9.64416	0.10369
40	45.25926	0.02209	442.59256	0.00226	9.77905	0.10226
45	72.89048	0.01372	718.90484	0.00139	9.86281	0.10139
50	117.39085	0.00852	1163.90853	0.00086	9.91481	0.10086

n	$(1+i)^n$	$(1+i)^{-n}$	$s_{\bar{n}}$	R'_n	$_nl_0$	R''_n
1	1.12000	0.89286	1.00000	1.00000	0.89286	1.12000
2	1.25440	0.79719	2.12000	0.47170	1.69005	0.59170
3	1.40493	0.71178	3.37440	0.29635	2.40183	0.41635
4	1.57352	0.63552	4.77933	0.20923	3.03735	0.32923
5	1.76234	0.56743	6.35285	0.15741	3.60478	0.27741
6	1.97382	0.50663	8.11519	0.12323	4.11141	0.24323
7	2.21068	0.45235	10.08901	0.09912	4.56376	0.21912
8	2.47596	0.40388	12.29969	0.08130	4.96764	0.20130
9	2.77308	0.36061	14.77566	0.06768	5.32825	0.18768
10	3.10585	0.32197	17.54874	0.05698	5.65022	0.17698
11	3.47855	0.28748	20.65458	0.04842	5.93770	0.16842
12	3.89598	0.25668	24.13313	0.04144	6.19437	0.16144
13	4.36349	0.22917	28.02911	0.03568	6.42355	0.15568
14	4.88711	0.20462	32.39260	0.03087	6.62817	0.15087
15	5.47357	0.18270	37.27971	0.02682	6.81086	0.14682
16	6.13039	0.16312	42.75328	0.02339	6.97399	0.14339
17	6.86604	0.14564	48.88367	0.02046	7.11963	0.14046
18	7.68997	0.13004	55.74971	0.01794	7.24967	0.13794
19	8.61276	0.11611	63.43968	0.01576	7.36578	0.13576
20	9.64629	0.10367	72.05244	0.01388	7.46944	0.13388
21	10.80385	0.09256	81.69874	0.01224	7.56200	0.13224
22	12.10031	0.08264	92.50258	0.01081	7.64465	0.13081
23	13.55235	0.07379	104.60289	0.00956	7.71843	0.12956
24	15.17863	0.06588	118.15524	0.00846	7.78432	0.12846
25	17.00006	0.05882	133.33387	0.00750	7.84314	0.12750
26	19.04007	0.05252	150.33393	0.00665	7.89566	0.12665
27	21.32488	0.04689	169.37401	0.00590	7.94255	0.12590
28	23.88387	0.04187	190.69889	0.00524	7.98442	0.12524
29	26.74993	0.03738	214.58275	0.00466	8.02181	0.12466
30	29.95992	0.03338	241.33268	0.00414	8.05518	0.12414
35	52.79962	0.01894	431.66350	0.00232	8.17550	0.12232
40	93.05097	0.01075	767.09142	0.00130	8.24378	0.12130
45	163.98760	0.00610	1358.23003	0.00074	8.28252	0.12074
50	289.00219	0.00346	2400.01825	0.00042	8.30450	0.12042

n	$(1+i)^n$	$(1+i)^{-n}$	$s_{\bar{n}}$	R_n'	$_n t_0$	R_n''
1	1.15000	0.86957	1.00000	1.00000	0.86957	1.15000
2	1.32250	0.75614	2.15000	0.46512	1.62571	0.61512
3	1.52088	0.65752	3.47250	0.28798	2.28323	0.43798
4	1.74901	0.57175	4.99338	0.20027	2.85498	0.35027
5	2.01136	0.49718	6.74238	0.14832	3.35216	0.29832
6	2.31306	0.43233	8.75374	0.11424	3.78448	0.26424
7	2.66002	0.37594	11.06680	0.09036	4.16042	0.24036
8	3.05902	0.32690	13.72682	0.07285	4.48732	0.22285
9	3.51788	0.28426	16.78584	0.05957	4.77158	0.20957
10	4.04556	0.24718	20.30372	0.04925	5.01877	0.19925
11	4.65239	0.21494	24.34928	0.04107	5.23371	0.19107
12	5.35025	0.18691	29.00167	0.03448	5.42062	0.18448
13	6.15279	0.16253	34.35192	0.02911	5.58315	0.17911
14	7.07571	0.14133	40.50471	0.02469	5.72448	0.17469
15	8.13706	0.12289	47.58041	0.02102	5.84737	0.17102
16	9.35762	0.10686	55.71747	0.01795	5.95423	0.16795
17	10.76126	0.09293	65.07509	0.01537	6.04716	0.16537
18	12.37545	0.08081	75.83636	0.01319	6.12797	0.16319
19	14.23177	0.07027	88.21181	0.01134	6.19823	0.16134
20	16.36654	0.06110	102.44358	0.00976	6.25933	0.15976
21	18.82152	0.05313	118.81012	0.00842	6.31246	0.15842
22	21.64475	0.04620	137.63164	0.00727	6.35866	0.15727
23	24.89146	0.04017	159.27638	0.00628	6.39884	0.15628
24	28.62518	0.03493	184.16784	0.00543	6.43377	0.15543
25	32.91895	0.03038	212.79302	0.00470	6.46415	0.15470
26	37.85680	0.02642	245.71197	0.00407	6.49056	0.15407
27	43.53531	0.02297	283.56877	0.00353	6.51353	0.15353
28	50.06561	0.01997	327.10408	0.00306	6.53351	0.15306
29	57.57545	0.01737	377.16969	0.00265	6.55088	0.15265
30	66.21177	0.01510	434.74515	0.00230	6.56598	0.15230
35	133.17552	0.00751	881.17016	0.00113	6.61661	0.15113
40	267.86355	0.00373	1779.09031	0.00056	6.64178	0.15056
45	538.76927	0.00186	3585.12846	0.00028	6.65429	0.15028
50	1083.65744	0.00092	7217.71628	0.00014	6.66051	0.15014

n	$(1+i)^n$	$(1+i)^{-n}$	$s_{\bar{n}}$	R'_n	${}_n t_0$	R''_n
1	1.20000	0.83333	1.00000	1.00000	0.83333	1.20000
2	1.44000	0.69444	2.20000	0.45455	1.52778	0.65455
3	1.72800	0.57870	3.64000	0.27473	2.10648	0.47473
4	2.07360	0.48225	5.36800	0.18629	2.58873	0.38629
5	2.48832	0.40188	7.44160	0.13438	2.99061	0.33438
6	2.98598	0.33490	9.92992	0.10071	3.32551	0.30071
7	3.58318	0.27908	12.91590	0.07742	3.60459	0.27742
8	4.29982	0.23257	16.49908	0.06061	3.83716	0.26061
9	5.15978	0.19381	20.79890	0.04808	4.03097	0.24808
10	6.19174	0.16151	25.95868	0.03852	4.19247	0.23852
11	7.43008	0.13459	32.15042	0.03110	4.32706	0.23110
12	8.91610	0.11216	39.58050	0.02526	4.43922	0.22526
13	10.69932	0.09346	48.49660	0.02062	4.53268	0.22062
14	12.83918	0.07789	59.19592	0.01689	4.61057	0.21689
15	15.40702	0.06491	72.03511	0.01388	4.67547	0.21388
16	18.48843	0.05409	87.44213	0.01144	4.72956	0.21144
17	22.18611	0.04507	105.93056	0.00944	4.77463	0.20944
18	26.62333	0.03756	128.11667	0.00781	4.81219	0.20781
19	31.94800	0.03130	154.74000	0.00646	4.84350	0.20646
20	38.33760	0.02608	186.68800	0.00536	4.86958	0.20536
21	46.00512	0.02174	225.02560	0.00444	4.89132	0.20444
22	55.20614	0.01811	271.03072	0.00369	4.90943	0.20369
23	66.24737	0.01509	326.23686	0.00307	4.92453	0.20307
24	79.49685	0.01258	392.48424	0.00255	4.93710	0.20255
25	95.39622	0.01048	471.98108	0.00212	4.94759	0.20212
26	114.47546	0.00874	567.37730	0.00176	4.95632	0.20176
27	137.37055	0.00728	681.85276	0.00147	4.96360	0.20147
28	164.84466	0.00607	819.22331	0.00122	4.96967	0.20122
29	197.81359	0.00506	984.06797	0.00102	4.97472	0.20102
30	237.37631	0.00421	1181.88157	0.00085	4.97894	0.20085
35	590.66823	0.00169	2948.34115	0.00034	4.99154	0.20034
40	1469.77157	0.00068	7343.85784	0.00014	4.99660	0.20014
45	3657.26199	0.00027	18281.30994	0.00005	4.99863	0.20005
50	9100.43815	0.00011	45497.19075	0.00002	4.99945	0.20002

APPENDIX C

TABLE C-1. AREA UNDER NORMAL PROBABILITY CURVE

Multiply values shown by 0.00001

z	.00	.01	.02	.03	.04	.05	.06	.07	.08	.09
0.0	00000	00399	00798	01197	01595	01994	02392	02790	03188	03586
0.1	03983	04380	04776	05172	05567	05962	06356	06749	07142	07535
0.2	07926	08317	08706	09095	09483	09871	10257	10642	11026	11409
0.3	11791	12172	12552	12930	13307	13683	14058	14431	14803	15173
0.4	15554	15910	16276	16640	17003	17364	17724	18082	18439	18793
0.5	19146	19497	19847	20194	20450	20884	21226	21566	21904	22240
0.6	22575	22907	23237	23565	23891	24215	24537	24857	25175	25490
0.7	25804	26115	26424	26730	27035	27337	27637	27935	28230	28524
0.8	28814	29103	29389	29673	29955	30234	30511	30785	31057	31327
0.9	31594	31859	32121	32381	32639	32894	33147	33398	33646	33891
1.0	34134	34375	34614	34850	35083	35313	35543	35769	35993	36214
1.1	36433	36650	36864	37076	37286	37493	37698	37900	38100	38298
1.2	38493	38686	38877	39065	39251	39435	39617	39796	39973	40147
1.3	40320	40490	40658	40824	40988	41149	41308	41466	41621	41774
1.4	41924	42073	42220	42364	42507	42647	42786	42922	43056	43189
1.5	43319	43448	43574	43699	43822	43943	44062	44179	44295	44408
1.6	44520	44630	44738	44845	44950	45053	45154	45254	45352	45449
1.7	45543	45637	45728	45818	45907	45994	46080	46164	46246	46327
1.8	46407	46485	46562	46638	46712	46784	46856	46926	46995	47062
1.9	47128	47193	47257	47320	47381	47441	47500	47558	47615	47670
2.0	47725	47778	47831	47882	47932	47982	48030	48077	48124	48169
2.1	48214	48257	48300	48341	48382	48422	48461	48500	48537	48574
2.2	48610	48645	48679	48713	48745	48778	48809	48840	48870	48899
2.3	48928	48956	48983	49010	49036	49061	49086	49111	49134	49158
2.4	49180	49202	49224	49245	49266	49286	49305	49324	49343	49361
2.5	49379	49396	49413	49430	49446	49461	49477	49492	49506	49520
2.6	49534	49547	49560	49573	49585	49598	49609	49621	49632	49643
2.7	49653	49664	49674	49683	49693	49702	49711	49720	49728	49736
2.8	49744	49752	49760	49767	49774	49781	49788	49795	49801	49807
2.9	49813	49819	49825	49831	49836	49841	49846	49851	49856	49861
3.0	49865									

BIBLIOGRAPHY

BOOKS

Allen, R. G. D.: "Mathematical Economics," 2d ed., St. Martin's Press, Inc., New York, 1959.

ARINC Research Corporation: "Reliability Engineering," Prentice-Hall, Inc., Englewood Cliffs, N.J., 1964.

Badger, Ralph E., Harold W. Torgerson, and Harry G. Guthmann: "Investment Principles and Practices," 6th ed., Prentice-Hall, Inc., Englewood Cliffs, N.J., 1969.

Barish, Norman N.: "Economic Analysis for Engineering and Managerial Decision-Making," McGraw-Hill Book Company, New York, 1962.

Baumol, William J.: "Economic Theory and Operations Analysis," 2d ed., Prentice-Hall, Inc., Englewood Cliffs, N.J., 1965.

Bierman, Harold, Jr., Charles P. Bonini, and Warren H. Hausman: "Quantitative Analysis for Business Decisions," 4th ed., Richard D. Irwin, Inc., Homewood, Ill., 1973.

Boulding, Kenneth E.: "Economic Analysis," 4th ed., Harper & Row, Inc., New York, 1966.

Bowker, Albert H., and Gerald J. Lieberman: "Engineering Statistics," 2d ed., Prentice-Hall, Inc., Englewood Cliffs, N.J., 1972.

Bursk, Edward C., and John F. Chapman: "New Decision-Making Tools for Managers," Harvard University Press, Cambridge, Mass., 1963.

Clendenin, John C., and George A. Christy: "Introduction to Investments," 5th ed., McGraw-Hill Book Company, New York, 1969.

Conard, Joseph W.: "Introduction to the Theory of Interest," University of California Press, Berkeley, Calif., 1959.

De Garmo, E. Paul: "Engineering Economy," 4th ed., The Macmillan Company, New York, 1967.

Dewhurst, R. F. J.: "Business Cost-Benefit Analysis," McGraw-Hill Book Company, New York, 1973.

Enrick, Norbert L.: "Quality Control and Reliability," 6th ed., Industrial Press, Inc., New York, 1972.

Financial Publishing Company: "Comprehensive Bond Values Tables," 4th ed., Boston, Mass., 1958.

————: "Financial Compound Interest and Annuity Tables," 5th ed., Boston, Mass., 1970.

Freund, John E.: "Modern Elementary Statistics," 3d ed., Prentice-Hall, Inc., Englewood Cliffs, N.J., 1967.

Giffin, Walter C.: "Introduction to Operations Engineering," Richard D. Irwin, Inc., Homewood, Ill., 1971.

Grant, Eugene L., and W. Grant Ireson: "Principles of Engineering Economy," 5th ed., The Ronald Press Company, New York, 1970.

——— and Richard S. Leavenworth: "Statistical Quality Control," 4th ed., McGraw-Hill Book Company, New York, 1972.

Havens, R. Murray, John S. Henderson, and Dale L. Cramer: "Economics: Principles of Income, Prices, and Growth," The Macmillan Company, New York, 1966.

Howard, Bion B., and Miller Upton: "Introduction to Business Finance," McGraw-Hill Book Company, New York, 1953.

Hummel, Paul M., and Charles L. Seebeck, Jr.: "Mathematics of Finance," 3d ed., McGraw-Hill Book Company, New York, 1971.

Ireson, W. Grant, and Eugene L. Grant: "Handbook of Industrial Engineering and Management," 2d ed., Prentice-Hall, Inc., Englewood Cliffs, N.J., 1971.

Kwak, N. K.: "Mathematical Programming with Business Applications," McGraw-Hill Book Company, New York, 1973.

Lindgren, B. W.: "Statistical Theory," 2d ed., The Macmillan Company, New York, 1968.

Loomba, N. Paul: "Linear Programming: An Introductory Analysis," McGraw-Hill Book Company, New York, 1964.

Murphy, James L.: "Introductory Econometrics," Richard D. Irwin, Inc., Homewood, Ill., 1973.

Musselman, Vernon A., and Eugene H. Hughes: "Introduction to Modern Business," 6th ed., Prentice-Hall, Inc., Englewood Cliffs, N.J., 1973.

Papoulis, Athanasios: "Probability, Random Variables, and Stochastic Processes," McGraw-Hill Book Company, New York, 1965.

Park, William R.: "Cost Engineering Analysis," John Wiley & Sons, Inc., New York, 1973.

Parks, R. D.: "Examination and Valuation of Mineral Property," 4th ed., Addison-Wesley Publishing Company, Reading, Mass., 1957.

Richmond, Samuel B.: "Operations Research for Management Decisions," The Ronald Press Company, New York, 1968.

Riggs, James L.: "Economic Decision Models for Engineers and Managers," McGraw-Hill Book Company, New York, 1968.

Schweyer, Herbert E.: "Process Engineering Economics," McGraw-Hill Book Company, New York, 1955.

Siddall, James N.: "Analytical Decision-Making in Engineering Design," Prentice-Hall, Inc., Englewood Cliffs, N.J., 1972.

Smith, Gerald W.: "Engineering Economy: Analysis of Capital Expenditures," 2d ed., Iowa State University Press, Ames, Iowa, 1973.

Spivey, W. Allen: "Linear Programming," The Macmillan Company, New York, 1963.

Stigler, George J.: "The Theory of Price," 3d ed., The Macmillan Company, New York, 1966.

Taha, Hamdy A.: "Operations Research: An Introduction," The Macmillan Company, New York, 1971.

Taylor, George A.: "Managerial and Engineering Economy," Van Nostrand Reinhold Company, New York, 1964.

Thuesen, H. G., W. J. Fabrycky, and G. J. Thuesen: "Engineering Economy," 4th ed., Prentice-Hall, Inc., Englewood Cliffs, N.J., 1971.

U.S. Department of Commerce, National Bureau of Standards: "Tables of Binomial Probability Distribution" (Applied Mathematics Series no. 6), 1950.

Volk, William: "Applied Statistics for Engineers," 2d ed., McGraw-Hill Book Company, New York, 1969.

Wagner, Harvey M.: "Principles of Operations Research," Prentice-Hall, Inc., Englewood Cliffs, N.J., 1969.

PERIODICALS

The American Economic Review, published five times a year by The American Economic Association.

Econometrica, published bimonthly by the Econometric Society.

The Engineering Economist, published quarterly by the American Society for Engineering Education, Engineering Economy Committee.

Industrial Engineering, published monthly by the American Institute of Industrial Engineers.

Management Science, published monthly by The Institute of Management Sciences.

Operations Research, published bimonthly by the Operations Research Society of America.

INDEX

INDEX

i = interest rate per period

n = number of interest periods

P = present value of a sum of money

F = the future value of a sum after n periods

A = a uniform end of period payment over n periods

future value $\quad F_n = P(1+i)^n$

present value of a future sum $\quad P = \dfrac{F}{(1+i)^n}$

future worth for series of year end payments, 1st payment 1 yr. from start, last payment at end

$$F = A\left[\frac{(1+i)^n - 1}{i}\right]$$

annuity (A) necessary to result in a future sum (F)

$$A = F\left[\frac{i}{(1+i)^n - 1}\right]$$